Start with a Scan

A Guide to Transforming Scanned Photos and Objects into High Quality Art

by

Janet Ashford and John Odam

PEACHPIT PRESS

Start with a Scan:
A Guide to Transforming Scanned Photos
and Objects into High Quality Art

Janet Ashford and John Odam

Peachpit Press
1249 Eighth Street
Berkeley, CA 94710
(800) 283-9444, (510) 524-2178
(510) 524-2221 (fax)

Find us on the World Wide Web at:
http://www.peachpit.com

Peachpit Press is a division of Addison Wesley
Longman.

DEDICATION

To Linnea Dayton, for many years of true and caring friendship.
—Janet Ashford

To my brother, George, and sister, Mary—my teachers in art, music and literature.
—John Odam

ACKNOWLEDGMENTS

We would like to thank the following people who served as photographic models: Florence and Molly Ashford, Hayden Lee Bird, Garrick Davis, Luisa Guilmon, Marc Hansen, Paul Henry, Abby and Alison Odam, Margaret Reid, Barry Shultz and Hale Thatcher. All original photographs used in this book were taken by Janet Ashford and John Odam except as otherwise noted in the captions. Thanks to Dave Allen, Florence Ashford, Meg Davis and Jennifer Church for permission to reproduce their photographic work and to Karl Nicholasen for permission to use his illustrations.

Microtek and LaCie provided helpful answers to our technical questions. Special thanks to Karen Luthart of Microtek and John Boehme of LPS Computer Service Group for hardware support and advice.

Thanks to the following organizations for permission to use graphics created for their multimedia presentations: America West, Cahners, the Colorado Rockies, the Gosney Company, IVI, the New Mexico Governor's office and Sir Speedy. Thanks also to Dru Jacobs and John Jensen for Web page development; and to the following publishers for permission to reproduce book covers and page designs: the California Energy Extension Service, Greenhaven Press, ITP and Pfeiffer.

Special thanks to Linnea Dayton for performing an excellent technical edit of our manuscript. Thanks also to David Heiret for his careful indexing and to Ted Nace and Nolan Hester of Peachpit Press for their enthusiastic support of our project. John McWade, editor of *Before & After* and Russell Brown of Adobe Systems reviewed our manuscript and providing helpful comments.

Many thanks to Doug Isaacs and Larry Christian of Adage Graphics in Los Angeles for their excellent service in producing film from our PageMaker files.

Contents

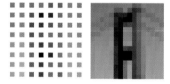

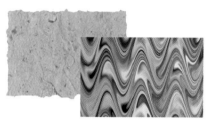

1 | Getting Started

Introduction

HOW WE CAME TO WRITE THIS BOOK

I have been doing graphic design and illustration for twenty-five years and writing "how-to" articles and books specifically on computer graphics for about six years now, honing the skill of describing in a step-by-step way how illustrators and graphic designers create their best work. Over the years I've noticed that almost every time I ask a computer artist how an image was created, the answer is "Well, I started with a scan of …" And when developing an illustration, I often start with a scan myself. So, the idea was born for a book that would show how to transform raw scans into good-looking electronic illustrations. I developed the original idea and outline for the book and then asked my long-time friend and colleague John Odam to help with the project, knowing that his talent and many years of experience as a graphic designer and illustrator would make the book stronger. We have divided the task, each contributing half of the text and artwork.

Throughout the process I've come to believe that the scanner is a *vital* tool for the computer artist because it brings *non*-computer-generated images (photographs, drawings, sketches, clip art, actual objects) onto the computer drawing board, so to speak, where they can be further manipulated and refined. Without this link to the *natural* world, electronic illustration risks being cold and slick. But when the rich textures and colors, and the irregular shapes and lines of traditional art—rough pencil sketches, pen-and-ink drawings, snapshots, old engravings, found art and so on—form the starting point, computer art has more warmth, charm, humor and depth, as well as touches of appealing quirkiness. In a word, it becomes more *human*. With this in mind, I hope you enjoy reading and using *Start with a Scan*.
— *Janet Ashford*

In the beginning
These Thunderscan images were edited in MacPaint *circa* 1986. Things were pretty jagged in those days.

My first exposure to computer scanning was at the home of an enthusiastic new Macintosh owner in 1985. He had a dot matrix printer that made a noise like a bee, industriously producing page after page of wicker-filled ovals.

Nice, but that's not all.

The ribbon on the printer could be replaced with a gadget called a Thunderscan that could actually read photographs that were fed into the printer.

Cool!

I marvelled at the flickering, grainy images that slowly emerged on the tiny blue screen and began to wonder if one might make a living with such toys.

In a couple of years or so, making a living with this stuff was, well, no problem. Strange gray boxes with dangling cables began to sprout around the studio. Now you could plug a video camera into your computer. Soon the flatbed scanner would let us see the world in many shades of gray. Layouts could include real photos. The gap between graphic design and desktop publishing began to narrow. Next, an explosion of color on the computer took the design community by storm. Even the doubters began to see that the graphic design industry was going to go digital. The revolution had arrived. There was no turning back.

Today, with a color scanner on almost every desktop, scanning has become a routine, mundane affair—no more miraculous than a fax or a modem. Scanners are just being taken for granted. So when Janet explained her idea to me I knew she was onto something. I knew that if we could rekindle that sense of wonder and excitement that occurred when we first started messing around with scans, and if we could temper it with years of practical experience, we'd have a book that would be both useful and fun. I hope you agree.
— *John Odam*

What happened to a
sense of wonder?
—Van Morrison

WHAT THIS BOOK IS ABOUT

Start with a Scan is a visual, step-by-step guide intended to show designers and illustrators how to transform raw scanned images into good-looking finished illustrations. The book includes three chapters on the technical basics of scanning, to provide the information needed to get images out of the computer and onto the printed page. But the bulk of the book is devoted to showing you how to start with a scan of almost anything (a lackluster photo, a clip art engraving, a household object) and use either image-editing or PostScript illustration software to turn it into an original, high-quality piece of art. Every chapter includes attractive color and black-and-white graphics, clearly written text, captions with how-to details, sidebars on special topics, and occasional quotes on related ideas, including creativity. We have tried to make the book visually seductive, easy to browse and read, well-organized, useful, and an inspiration to designers, illustrators and students. With this book in hand you'll be able to:

• Learn how to start with a scan of *anything* and turn it into high-quality art.

• Browse through hundreds of full-color illustrations for ideas you can use.

• Follow clear, step-by-step instructions to get the results you want.

IS THIS BOOK FOR YOU?

Although *Start with a Scan* will be useful for anyone who owns a scanner or is planning to get one, it is especially directed to those who use personal computers to create pages for publication, either on paper or in the electronic media, professionally or just for fun.

Throughout our planning we've had two particular groups of readers in mind:

• first, experienced designers and illustrators who

Hold still, please!
Early experiments with MacVision using a surveillance video camera, tripod and tangle of cables produced these results.

Adding color
This video portrait was saved in MacVision's "zebra" mode and then colored with an early version of PixelPaint, *circa* 1988.

work with computer graphics every day and are always looking for new ideas and techniques to stimulate their imaginations and broaden their repertoire,

• and second, design students and people who are new to the computer graphics field, who want to know what can be done in this medium.

If you have a love of visual images and a desire to tinker with them, this book's for you.

BACKGROUND AND SOFTWARE YOU WILL NEED

We assume that our professional readers are familiar with the most popular desktop graphics programs, including Adobe Photoshop, a PostScript drawing program (such as Macromedia FreeHand or Adobe Illustrator), an autotracing program (such as Adobe Streamline), and a page layout program (such as QuarkXPress or Adobe PageMaker), on either a Macintosh or an IBM-type platform. With this background, you should be able to go directly from the book to your own computer to duplicate the effects you see on these pages. We don't include click-and-drag instructions for every technique because we want the book to be applicable for a variety of different software programs and computers. For the same reason, screen dumps and menus are rare. But we do include enough detail so that you can translate from the system we're using to the one you're using. We do rely on Photoshop as our workhorse image-editor and we refer to Photoshop-specific features where necessary to clarify a procedure, knowing that other image-editing programs include similar functions.

How To Use This Book

LEARNING BY LOOKING

Start with a Scan is first and foremost a visual stimulus. Though we have worked to make the text as clear and concise as possible, it should be possible to get a lot of information from this book simply by looking at the artwork, without reading a single word.

So first of all, look at it! *Start with a Scan* is packed with color illustrations designed to make you keep turning the pages, to get you excited about all the interesting things that can be done with a scan, and to encourage you to try out our techniques on your own computer. The book is meant to stimulate your thinking about design and illustration, and about how you can expand the use of your scanner and computer as creative tools. Like some computer systems, the creative process often "hangs." So keep this book handy when you're working out new ideas—it may keep you from getting stuck. When you see something you particularly like and want to try, just read the caption next to the image. It will describe the technique in a quick, simple way that should get you started. For more details, and for background on the images and techniques, also read the running text.

ORGANIZING THINGS

We have divided the book into sections organized around the different ways scans can be used to generate original graphics for a wide range of media—from billboards to boxtops. If you are new to scanning you would be well-advised to begin with chapters 2, 3 and 4, which describe the basics of scanners, scanning and image-editing. Keep in mind though that this is not a book about scanning technique *per se*. Other books do an admirable job of explaining the technical aspects of scanning, including *Real World Scanning and Halftones* by David Blatner and Steve Roth (Peachpit Press, 1993),

which covers technical issues in depth. If you are already familiar with the basics, we invite you to explore these pages in any order. For the most part, each page or spread is a self-contained explanation of a particular technique or resource.

We have placed the original black-and-white scan from which an image was made over a tan box so that it can be easily identified. We have also placed each piece of finished art (such as book covers, product labels, posters and so on) over a drop shadow that makes it easy to distinguish from the step-by-step figures. Sidebars are also placed over tan backgrounds and are self-contained small lessons on special topics.

HOW THIS BOOK WAS PRODUCED

The illustrations in this book were created in Adobe Photoshop, Fractal Design Painter, Adobe Illustrator, Macromedia FreeHand, Altsys Fontographer, Adobe Dimensions and Strata StudioPro. The book was designed and laid out in Adobe PageMaker. Photoshop was used for all image-editing. We used Macintosh computers, but all the techniques shown here can also be used with the same programs running on Windows, and most can be used with other image-editing or drawing programs.

SECTION TITLE

FINISHED ART DEVELOPED FROM SCANS

SIDEBAR ON A SPECIAL TOPIC

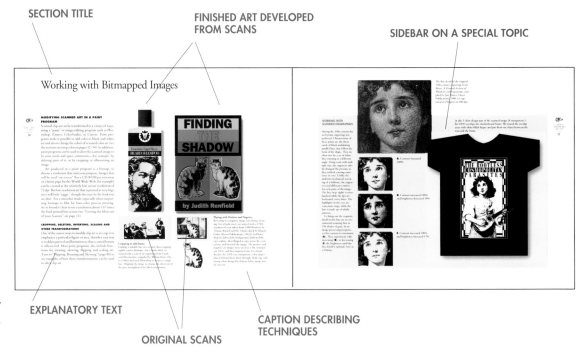

EXPLANATORY TEXT

ORIGINAL SCANS

CAPTION DESCRIBING TECHNIQUES

2 | Working with Scanners

Scanners and Digitizing Devices
Types of Scanners
Digital Cameras, Still Video, Video Frame Grabbers

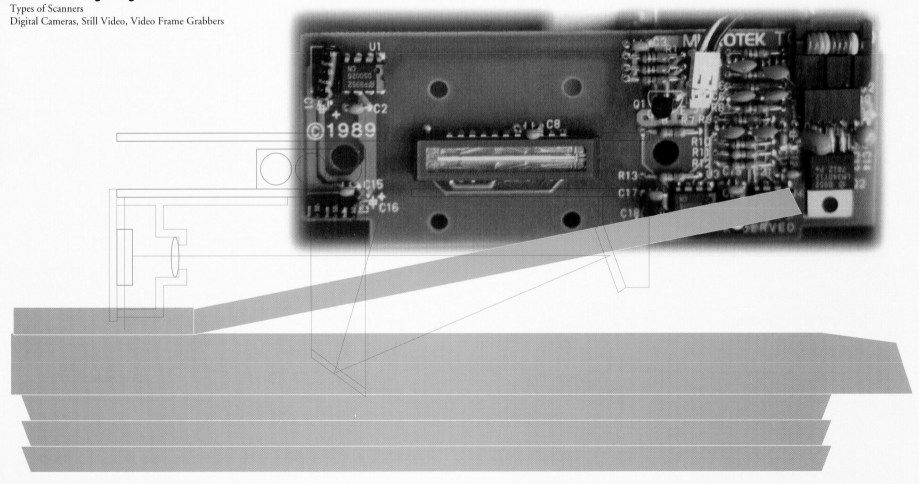

Scanners and Digitizing Devices

TYPES OF SCANNERS

Ten short years ago, we nearly bought a photostat camera for copying graphics. It was secondhand and it was a deal. And the cost of buying out veloxes and prints of line art was not trivial for a small design studio. The main drawback was that a stat camera and its processing unit would take up a whole room. So we invested instead in a Macintosh computer, without realizing that soon this curious hybrid of a TV set and a typewriter would be hooked up to a scanning device far more powerful and versatile than a room-filling stat camera (and a lot cheaper, too).

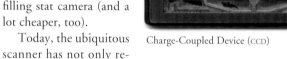

Charge-Coupled Device (CCD)

Today, the ubiquitous scanner has not only re-placed the old stat camera, but has also made remarkable inroads into the color separation industry. Here are some of the main types.

DRUM

The drum scanner is the typical "high-end" equipment used by color separators who make the film used for printing. It uses a *photomultiplier tube* (PMT) as a sensor to capture the image. PMTs accurately measure the light and dark value of an image line-by-line vertically as the drum rapidly revolves. Filters are used to extract the color information.

Drum scanners can cost as much as $50,000. Most use specialized software and hardware and are geared toward the maximum quality of resolution and color fidelity. Recently, lower-cost, scaled-down versions of drum scanners have been introduced for the desktop market.

FLATBED

The typical workhorse scanner in desktop publishing is the flatbed. Ranging from $500 to $2,000, most operate at 1-bit (line art), 8-bit grayscale (256 grays), and 24-bit color (over 16 million colors). Flatbed scanners use an array of *charge-coupled devices* (CCDs) to convert the image to digital information, reading the image in a series of horizontal strips.

Two basic flatbed scanner types are currently available: three-pass and one-pass. A *three-pass scanner* uses one light source and three filters to generate the RGB (red, green, blue) values needed for a color scan; the light source travels across the original three times. *One-pass scanners* use three separate strobe lights (red, green and blue), that flash alternately in rapid succession as the scanning mechanism passes once across the original. Although faster than the three-pass, they can cause multicolored shadows on scans of originals that are not absolutely flat.

Some flatbed scanners come with attachments that allow slides and transparencies to be scanned. The quality of these slide scans tends to be low, however, and suitable for position only (FPO) in layout work.

SLIDE SCANNER

Dedicated slide scanners are available in the $1,000 to $3,000 range and are mostly limited to the 35 mm format. They are particularly useful to commercial photographers who like to provide their clients with digital images, or who use their slides to create photo montages on the computer. Some slide scanners are designed for internal installation in the computer. Although the image quality from a slide scanner is good, its quality is not quite as high as a drum scanner.

An economical alternative to owning a slide scanner is to have your slides or color negatives digitized onto a Kodak Photo CD disk at cost of $1 to $3 per image. For

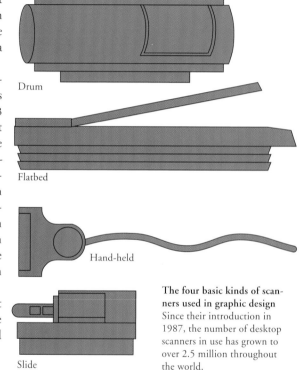

Drum

Flatbed

Hand-held

Slide

The four basic kinds of scanners used in graphic design Since their introduction in 1987, the number of desktop scanners in use has grown to over 2.5 million throughout the world.

HOW A CHARGE-COUPLED DEVICE WORKS

A charge-coupled device (CCD) is an array of tiny elements—up to 3,000 mounted on a chip in three rows (**A**). The function of the CCD is to measure the intensity of light reflected from thousands of small areas of the original and convert these measurements to digital information. In a three-pass scanner, the CCD has a single array of elements, as shown in the diagram. A one-pass scanner has three arrays coated with red, green and blue.

In this detail of the CCD array (**B**), the central row of silicon elements is light-sensitive. Light striking an element generates a negative charge proportional to the amount of light it receives. The two outside rows of elements carry the charge away from the central elements at regular intervals, resetting the light-sensitive elements to neutral. These elements read out the charge as digital pulses that are transmitted to the surrounding circuitry. This light-sampling process is rather like having a row of buckets in a field during a rainstorm, where the water level in the buckets is measured at regular intervals and the buckets are emptied to begin again.

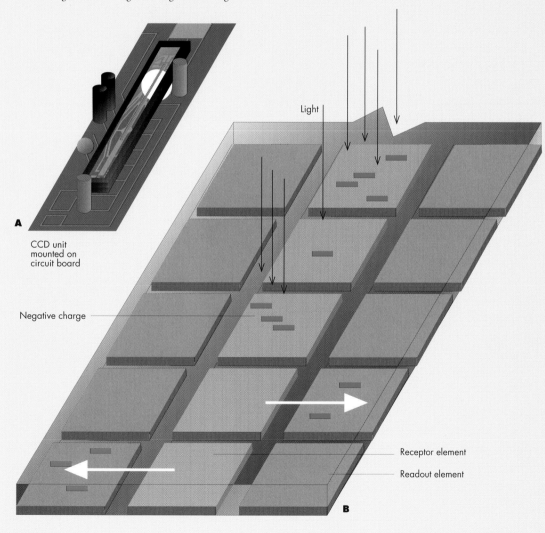

A

CCD unit
mounted on
circuit board

Negative charge

Light

Receptor element

Readout element

B

quality rivaling the drum scanner it might be worth waiting over a week for the processing.

HAND-HELD

Hand-held scanners cost about $200, are limited to originals up to four inches wide, and require a very steady hand. Cheap, but limited in capability, they are useful for scanning spot art from old and fragile books whose binding cannot withstand flattening out on a flatbed.

DIGITAL CAMERAS, STILL VIDEO, VIDEO FRAME GRABBERS

Other ways of capturing visual information digitally do not involve scanning.

Video frame grabbers are usually a combination of software and hardware that can acquire sequences or individual frames from video tape. These images can be incorporated into page layouts or multimedia projects. The hardware is typically a card that plugs into the motherboard of the computer. Some computers, however, come with built-in video image-processing chips. Video images are limited to 560 × 480 pixels resolution, and do not compare well to scanned photographs when printed.

Still video involves a portable camera similar to a regular 35 mm camera. Images are sensed by a video recording tube and stored on small reusable floppy disks that have a capacity of up to 100 exposures. The image quality is comparable to a video frame grab.

Digital cameras use CCD sensors and a variety of storage media. In a studio setting, the data can be fed directly to the computer and stored on a hard disk. These rather expensive cameras produce high-quality images equivalent to those made by slide scanners. They are ideal for catalog photography since the digital files can go directly into the layout.

HOW A SCANNER WORKS

When you click the mouse to start scanning, here's what happens:

(A) Commands from the computer instruct the logic board (**1**) to regulate the motor speed controller (**2**) and motor (**3**). The motor drives the transmission belt (**4**), which is connected to the scanning unit (**5**).

These logic board instructions move the scanning unit into the correct position to begin the scan; they also govern the speed of the motor during the scan. Light from the lamps (**6**) strikes the original artwork or photograph placed facedown on the glass top and is reflected from the mirrors (**7**) through the lens (**8**) and onto the CCD sensors (**9**).

Output from the CCD sensors is interpreted by the logic board and transmitted back to the computer.

(**B**) A section through the scanning unit that travels under the glass top shows two mirrors angled so that the light reflected from the original passes through a lens and is focused on the CCD sensors.

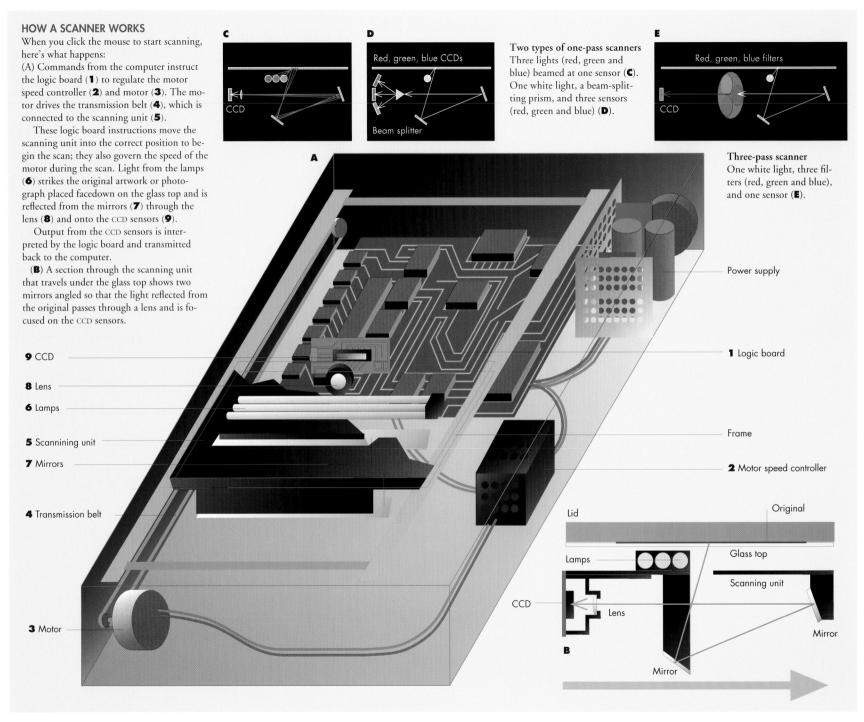

C

CCD

D

Red, green, blue CCDs

Beam splitter

Two types of one-pass scanners
Three lights (red, green and blue) beamed at one sensor (**C**). One white light, a beam-splitting prism, and three sensors (red, green and blue) (**D**).

E

Red, green, blue filters

CCD

Three-pass scanner
One white light, three filters (red, green and blue), and one sensor (**E**).

A

Power supply

1 Logic board

Frame

2 Motor speed controller

9 CCD

8 Lens

6 Lamps

5 Scannining unit

7 Mirrors

4 Transmission belt

3 Motor

Lid

Original

Lamps

Glass top

Scanning unit

CCD

Lens

Mirror

Mirror

B

3 | Technical Considerations

Planning Ahead

Interpolation
A detail of a scan (**D**) reveals a given amount of information. A scan at half that resolution (**E**) captures half as much information. Interpolation (**F**) spreads the pixels apart, and fills in the missing information (**G**) by averaging the differences between neighboring pixels.

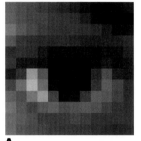

A

B

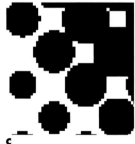

C

In a scan (**A**) the image is made of a grid of pixels of varying shades of gray, whereas in the printer's halftone (**B**) the picture is made of round dots that are all absolutely black. When the halftone dots are magnified further (**C**) it can be seen that the dots themselves are made up of smaller, square dots.

HOW WILL YOUR SCAN BE USED?

Scanning is the first step in a chain of events that leads to producing an image on some type of media. Whether the medium is print or on-screen display—laser printer, four-color offset printing, World Wide Web page, or CD-ROM—knowing how your scan will be used is essential in determining how that first step is made.

Most desktop equipment can't record the amount of information needed for full-page advertisements, large-format books, calendars, posters, or any other case where the image size is greater than 8 × 10 inches. High-quality offset color printing in larger sizes requires scans from commercial color separators.

Nevertheless, a desktop scanner *can* produce excellent results for offset printing in all but the most exacting cases. For example, if the original is a regular (continuous-tone) print, color or black-and-white, 8 × 10 inches or less, and your output size is 100% or less of the original size, you can make good-quality images suitable for offset printing. Desktop scanners are, of course, perfect for multimedia presentations, preliminary design work, and output to any kind of laser printer, all of which have less stringent resolution requirements.

FROM SCANNER TO PRINTED PAGE

Pixels—short for *picture elements*—are the "atoms" that comprise computer images. The information from a scanned image is digitized into a matrix of pixels, displayed to the monitor and stored in the computer's memory. To make a printed image, that information has to be interpreted by the *raster image processor* (RIP) in an output device and converted to black dots on paper or on the film used for making printing plates.

The resolution or fineness of a scanned image is determined by the number of *pixels per inch* (ppi), also known as *dots per inch* (dpi). In a scanned image the visual information is conveyed by a grid of pixels, which form a mosaic of adjacent squares of different colors or gray tones. By contrast, a printed image is usually made up of a halftone screen composed of a mesh of tiny dots that vary in size (darker areas are produced by larger dots.) It takes halftone screens in four colors—cyan, magenta, yellow and black—to make a printed color image. The fineness of a halftone screen is determined by the number of *lines* or rows of dots *per inch* (lpi). When images are output as halftone dots through an imagesetter, the halftone dots themselves are made up of tiny imagesetter dots. The number of imagesetter dots that can make up the larger halftone dot limits the number of different possible dot sizes, restricting the number of gray tones that can be reproduced. The finer the mesh of the screen, or the greater the number of lines per inch, the fewer the possible gray tones. This is why you may notice posterization or "banding" in low-resolution laser prints where the halftone screen has been set to a high number.

OPTICAL RESOLUTION

The *optical resolution* of a scanner is determined by the density of its array of CCDs. You can check the optical resolution of your scanner by reading the manual or contacting the manufacturer. Optical resolution fixes an absolute limit on the amount of information the scanner can extract from an original.

INTERPOLATION

Most scanning software allows scanning at up to four times the built-in optical resolution of the scanner. This is achieved by *interpolation*, a process that spreads out the information collected by the CCDs over a larger area and fills in the missing pixels by averaging. This leads to one of the Great Mysteries of Scanning: One might

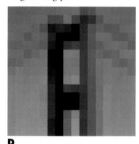

D

E

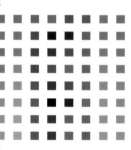

F

G

suppose that there is nothing to be gained by exceeding the optical resolution of your scanner, but in reality, we find image quality *can* be improved by scanning above optical resolution (see sidebar on this page).

GETTING THE MOST OUT OF YOUR SCANNER

All scanning processes require software to integrate the scanning hardware with the rest of your computer system. There are basically three software options: using the software that comes with the scanner, using a "plug-in" module that links your scanner to an image-processing program, or using third-party software that is designed to work with many different scanners and includes color-correction, automatic exposure and other controls. No matter which option you use, you will be able to specify resolution via the software before the scan is made. Typical settings range from 50 to 2,400 ppi.

Scanning at the highest possible resolution is not a good idea since it often wastes time and disk space, and produces more digital information than you need. The ideal to aim for is "optimal resolution," in which the scan resolution provides just as much information as the output and printing processes can effectively use, and no more.

OPTIMAL RESOLUTION FOR HALFTONE OUTPUT

The four main factors that affect the quality of an image printed in halftone are:

The number of pixels per inch in the scan (ppi)
The number of lines per inch in the halftone (lpi)
The resolution of the imagesetter (dpi)
The scale of the final image (%)

The process of converting square pixels into round dots requires that there be roughly two pixels of image information for every printed halftone dot, so it's

EXCEEDING OPTICAL RESOLUTION: IS IT WORTHWHILE?

When you scan at settings above the optical resolution of your scanner, the scanner's software fills in the missing pixels by interpolation. The picture below was scanned at three different resolutions on a LaCie Silverscanner: 300, 600 and 1,200 ppi. The scanner's optical resolution was 300

300 ppi

600 ppi (2 × optical)

1200 ppi (4 × optical)

ppi. The insets are enlarged 800%.

Scanning at higher resolutions takes much longer than scanning at optical resolution. So if an image needs to be enlarged, wouldn't it make more sense to scan quickly at optical resolution, then enlarge it later in an image processing program?

To find out, we took the same scan at 300 ppi and 600 ppi from the left column and used Adobe Photoshop with Bicubic interpolation to remap the picture up to 1,200 ppi., shown in the right column (below).

Our conclusion: The scanning software's interpolation

300 ppi enlarged to 1200 ppi

600 ppi enlarged to 1200 ppi

process does a better job than the post-production method.

Whether it's worthwhile to scan at above the optical resolution depends, of course, on the kind of original and how the scan will be used. In many

MEG DAVIS

300 ppi

600 ppi

situations little is to be gained by exceeding optical resolution. We scanned the picture of the brides-maids at optical and again at twice optical resolution. Can you tell the difference?

Line art, however, seems to benefit dramatically from interpolated scanning at a high ppi setting. In the example below you can compare two scans at optical and four times optical resolution.

300 ppi

1200 ppi

usually recommended that the pixels per inch (ppi) of the scan should be twice the lines per inch of the halftone (lpi). We have found, however, that a ratio of 1.67 ppi/lpi gives optimal resolution. So, for example, if your image will be printed with a 133 lpi screen, it should be scanned at 222 ppi (133 × 1.67). We practice what we preach. All the images in this

book, except line art, have been scanned at 250 ppi and printed at 150 lpi, a ratio of exactly 1.67 ppi/lpi.

SCALE AND RESOLUTION

Scanning at 100 ppi at 200% produces the same number of pixels as scanning at 200 ppi at 100%. The difference will be the dimensions and ppi of the im-

Scaling a scanned image
Left to right: 72 ppi at 100%;
72 ppi at 33%, equivalent to
216 ppi; 72 ppi at 300%,
equivalent to 24 ppi.

Resolution and quality
To demonstrate the relationship between half-tone screen frequency, scanning resolution and overall quality, we have shown the same image at six different resolutions from left to right—50, 100, 150, 225, 250 and 300 ppi, and at three different halftone screen frequencies from top to bottom—150, 100 and 50 lpi.

age when it's imported into a layout, drawing or image-editing program. Scaling the image up or down after it has been placed in a layout program also affects the resolution. A 72 ppi image scaled down to 33% in a layout program has an effective resolution of 216 ppi, but if it's scaled up 300%, its resolution will be only 24 ppi.

A formula for calculating the optimal resolution for a scan that also takes scaling into account is:

$$\frac{\text{Final image width}}{\text{Original image width}} \times \text{lpi} \times 1.67$$

RESOLUTION FACTORS IN LINE ART

Although image interpolation increases the resolution of images beyond the limit of optical resolution, for color or grayscale modes this will not usually increase the printed image quality.

Line art, on the other hand, can benefit from the use of image interpolation. The process smooths out the jagged edges of the image. Since line art files are small compared to grayscale or color files, there is no reason not to maximize the pixels per inch to get the best-quality output. If the line art original has a lot of fine detail, scan first in grayscale mode and use your image processing software to adjust the brightness and contrast to optimize the detail before converting the file to bitmap mode (see page 31).

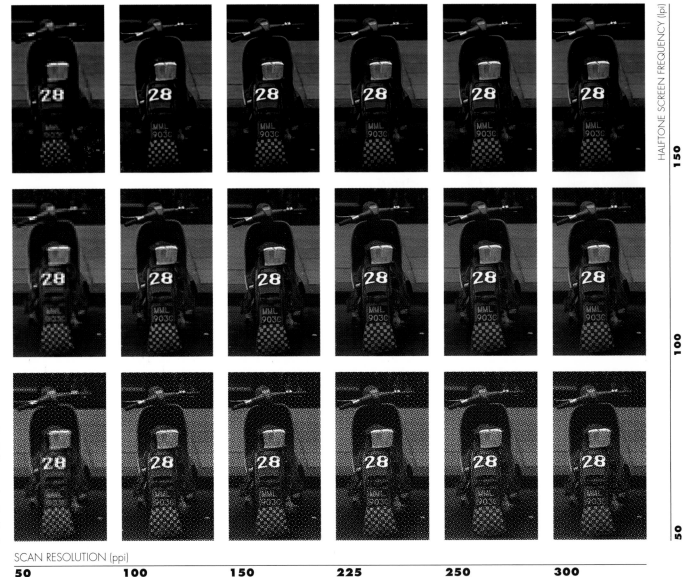

HALFTONE SCREEN FREQUENCY (lpi)

150

100

50

SCAN RESOLUTION (ppi)

| 50 | 100 | 150 | 225 | 250 | 300 |

Preparation, Formats and Modes

Image formats
From left to right: 24-bit color, 8-bit color, 256 grayscale.

SCANNING IN GRAYSCALE OR CONVERTING FROM COLOR: DOES IT MAKE A DIFFERENCE?

When you need a black-and-white scan of a color original you can either scan it in grayscale or scan it in color and convert it to grayscale later in an image-editing program.

The first method is quicker; the second method might improve image quality, depending on your scanner and on the conversion capabilities of your image-editing software. We recommend testing both methods.

MEG DAVIS

Grayscale scan (LaCie Silverscanner)

Grayscale scan (Microtek ScanMaker IIHR)

Color scan converted in Adobe Photoshop

Color scan converted in Adobe Photoshop

PREPARING THE ORIGINAL

An ideal original is a continuous-tone photographic print, no bigger than 8.5 × 11 inches, on opaque paper. Originals that are already halftones pose special problems: moiré and copyright (see pages 23–24). If the original is on thin paper with something printed on the back, there may be show-through that will be picked up by the scanner. The easy solution is to place a sheet of black paper behind the original and adjust the brightness if necessary.

GETTING IT STRAIGHT

Some scanning software automatically compensates for crooked originals. It can save time, however, if your original is already straight. Butting the edge of the original to the edge of the glass will ensure a straight scan. Close the lid *slowly* so that air pressure doesn't blow your original off kilter. If the original is out of square—for example, a smaller image taped casually to a larger piece of paper—use a T-square, cutting board and X-Acto knife to trim the edge of the larger paper in alignment with the taped image.

SCANNING MODES

Most scanning software provides for these scanning options: 16 million colors, 256 colors, 256 shades of gray, 16 shades of gray, line and halftone. The key options are 16 million colors and 256 shades of gray. The others may all be ignored, and some should be avoided altogether.

If the source image is color, always scan it at 16 million colors, even if your monitor or printer can only support 256 colors. A full-color scan will capture all the color information, always leaving you the option of contracting the color palette down to 256 colors later. We do not recommend *ever* scanning in the 256 color

mode, because the scanner has to take the extra time to pre-scan the image and calculate the best way to allocate the 256 available colors. Not only does it take twice as long to perform the scanning operation, but the resultant scan doesn't look as good, and the file cannot be successfully edited in an image-processing program, nor can it be converted back to 16 million colors.

If your source image is a black-and-white print or line art, consider using 256 shades of gray. If you want 16 shades of gray or fewer for a posterized effect, you will have more control by editing the file later rather than scanning at 16 shades. Line art can be converted to bitmap from grayscale after adjusting the brightness and contrast to eliminate noise and sharpen the detail. You don't have this kind of flexibility with a line scan.

The halftone mode is intended for low-end, dot-matrix or similar printers that cannot generate their own

STRAIGHTENING YOUR SCAN IN PREVIEW MODE

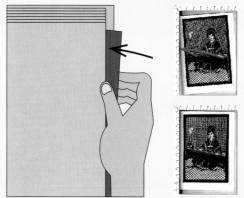

Because originals aren't always straight on the printed page, you may have to use the preview mode of your scanning software to check alignment. Straightening a scan in preview mode is counter intuitive. Move the original in the *same* direction that it appears to be leaning.

This original medical illustration by Bob Kinyon measures 17 × 15 inches, but the maximum image area of the flatbed scanner is 8.5 × 11.7 inches. The scanning was done in four overlapping tiles. The top two sections used the maximum image area of the scanner, but the lower sections only required a horizontal portion to ensure sufficient overlap. The amount of overlap need not be exact, but alignment is crucial. We used the edges of the art board to get the image straight.

The assembly process is shown below in sequence (**A–C**). We extended the canvas size of the first tile, placing it in the upper left corner. Each of the other tiles was copied and pasted carefully into position. We zoomed in to check alignment and used the nudge keys for final positioning. Cropping eliminated stepped edges (**D**). For another method of handling large originals see page 112.

halftones to reproduce grayscale images. It produces terrible-looking scans and should be avoided.

FILE FORMATS

Completed scans need to be saved at some point, but which file format is the best? The answer will depend on how the scan is going to be used and what operating system you are using, but the general strategy is to save in the most universal format for all possible applications and platforms.

TIFFS

TIFF, the acronym for *tagged image file format*, is the most widely used image format. TIFF files can be read by all image-editing programs, and they can be placed in all layout programs. Many drawing programs support TIFF files either as templates for tracing or for incorporation into the artwork. TIFFs will print out successfully from most programs. A TIFF file can be color, grayscale or line art. Grayscale images can be converted to line art by dithering, an effect akin to a mezzotint. Color files are RGB from the scanner, but may need to be converted to CMYK, depending on your printing requirements.

PICTS, GIFS AND JPEG

PICT files are widely used in multimedia and animation. Color files for multimedia should generally be 8-bit. On-line images for the World Wide Web should be saved as 8-bit GIF files, or as compressed JPEG files.

HANDLING LARGE ORIGINALS

Don't worry if your original is bigger than the scanning window. The limiting factors for image size are computer memory and human patience, not the size of your scanner. Tiling the scan and welding the pieces together in your image-processing software is relatively easy, provided that you allow some overlap and that initial scans are straight.

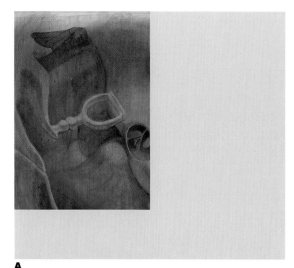

A

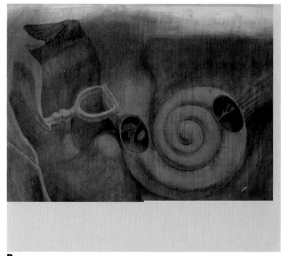

B

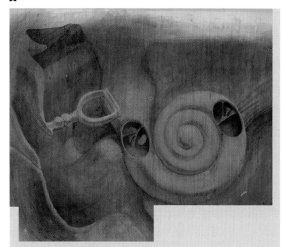

C

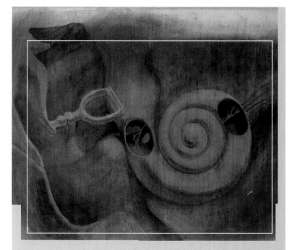

D

4 | Editing Scanned Images

Working with Tone and Color

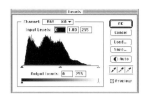

A Original image with corresponding Levels histogram

IMPROVING IMAGE QUALITY

Once you've made a scan at the appropriate resolution for your intended use, it's time to open it in an image-editing program and check the quality of the image: Are there dust specks or scratches? Is the picture well-focused? Is the color balance accurate and pleasing? The sequencing of any fix-up work is often important. In general, it's best to correct tonal range and color balance first, for if these cannot be improved, the image may have to be discarded. Sharpen the image next and then go on to make necessary changes in the content of the image, for example, repairing torn areas, cropping to improve the composition, and so on. Most photographs, no matter how poorly exposed, out of focus, or badly composed, can be significantly improved with electronic darkroom tools. For image-editing we use Photoshop, and that program's specific functions and tools are referred to in this chapter. Similar tools are available in other image-editing programs, including Painter, Collage and PhotoStyler.

CALIBRATION

Before you attempt any corrections to tonal range and color balance, it's important that your system be calibrated so that there is a predictable relationship between what you see on-screen and what you get when your art is output to paper or film. The "system" includes your monitor, scanner, scanning and image-editing software, and output device (imagesetter). Refer to the materials that came with your system components for information on calibration.

IMPROVING TONAL RANGE

Tonal range refers to the range of light and dark areas in an image and is a concern for both black-and-white and color scans. Unless the subject is something like a ski slope or a cave, an image should have a full range of midtone values between white and black. But note that prepress professionals suggest that the end points of the tonal scale should not be pure black or white but should set at percentages of, for example, 3% black for white and 97% black for black (see "Dot Gain: The Screen Lies" on page 17). In color images, adjustments to tonal range are usually made to the full RGB image, but they can also be made to each color channel individually if you are trying to remove a color cast, for example.

BRIGHTNESS AND CONTRAST

When faced with a muddy picture, it seems easiest to reach for the Brightness and Contrast controls. But these only shift the entire image up or down in brightness, without changing the relationship between the two extremes and the midtones. So in trying to lighten or darken the midtones you may blow out the highlights or plug up the shadows. Using histogram controls (Levels in Photoshop) or a Gamma curve (Curves in Photoshop) makes it possible to adjust the brightness of the midtones only and provides greater control over editing.

WHAT IS A HISTOGRAM?

A "histogram" is a graph display that plots the dark to light values of a continuous-tone image along the *x,* or horizontal, axis and the number of pixels found at each lightness value on the *y,* or vertical, axis. Whenever an image is open in Photoshop, the Levels command provides a dialog box with a histogram (labeled "Input") which displays the tonal range of the image. The Levels command can be used to set the black and white points and also to edit the midtones without significantly affecting the shadows and highlights.

SETTING THE BLACK AND WHITE POINTS

Defining values for the darkest and lightest pixels in an image is the first tonal range adjustment to make. This is called "setting black and white points." In a given image, "black" and "white" might have values of 10 and 240 respectively, over a total of 256 possible levels of gray (or brightness) from 0 to 255. Resetting the black and white points to 0 and 255 will usually improve contrast by spreading the brightness values of the pixels over a broader range and can be done automatically by clicking the Auto button in the Levels dialog box.

Adjustments to black and white points can also be made manually. Pulling the black and white sliders inward on the Levels Input scale (thus shortening the range of the input image) remaps those values out to 0 and 255 on the output scale, which increases contrast. On the other hand, pulling the black and white sliders inward on the Output scale reduces contrast by re-mapping the 0 and 255 levels inward (for example, to 20 and 230) so that the tonal range of the image is reduced.

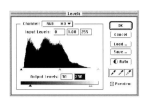

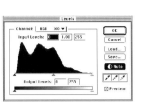

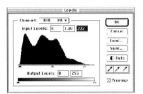

B Increasing contrast through Brightness/Contrast controls

C Increasing contrast by clicking Auto to reset black and white points.

D Increasing contrast by moving Input black and white sliders inward

E Decreasing contrast by moving Output black and white sliders inward

Histogram after using Brightness/contrast controls

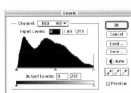

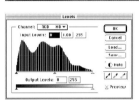

Histogram after using Auto button to reset black and white points

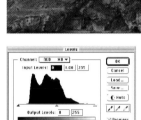

Histogram after using black and white Input sliders to reset black and white points

Histogram after reducing contrast by using black and white Output sliders

Improving contrast

A photo of a castle on a lake in Northern Italy was taken on a cloudy day and lacks contrast. The histogram for the image, displayed in Photoshop's Levels dialog box, shows a lack of pixels at the extreme ends of black and white, and generally fewer pixels in the light end of the midrange (**A**). The easiest, most intuitive way to increase contrast in an image like this is to simply to increase the contrast in the Contrast/Brightness dialog box. But doing so produces an image in which the midrange values are flattened and lacking in detail (**B**). A more effective way to improve contrast is to increase the width of the tonal range of the image by resetting the black and white points to 0 and 255. This spreads all the gray values over the full range of 256 possible gray or brightness levels. Resetting the black and white points can be done automatically by clicking on the Auto button in Photoshop's Levels dialog box. The resetting produces an increase in contrast in the midtones without dramatically increasing the brightness of the highlights or the darkness of the shadows (**C**). The black and white points can also be set "manually" by moving the black and white sliders on the Input scale in the Levels dialog box inward toward the center, which reduces the tonal range of the original image and remaps it outward to the full output range of 256 gray levels, thus in effect resetting the black and white points at the values you choose. In the castle image, we moved the black and white inward just to the edges of the spread of pixels. This produced an image with improved contrast and good detail in the midtones (**D**). It's also possible to reduce contrast by moving the black and white sliders on the Output scale toward the center. This reduces the tonal range of the image, producing an even flatter image of the castle than the original (**E**). This procedure could be used with an original that had too much contrast.

DOT GAIN: THE SCREEN LIES

Dot gain is what happens when the ink in a halftone dot spreads out when it's printed on paper. The ink will spread only slightly on coated paper, more on uncoated paper and even more on newsprint. Unfortunately, this means that even though you can adjust your image to look terrific on the screen, it will probably print too dark, shadow detail will be lost and the midtones may also look muddy.

To overcome dot gain problems we recommend first making adjustments to the file to optimize the image on screen. Once you're satisfied, save it and then prepare a duplicate version that will be corrected for printing. If you are working in Photoshop, the adjustments should be made in the Input/Output Levels controls in the Levels menu. Ask your printer how much dot gain correction you should use and whether to adjust for dot gain in the midtones or in the shadows—or both.

To make dot gain corrections in the midtones, enter a value in the middle box next to the words "Input levels." For example, entering 1.35 would make a 10% reduction and entering 1.95 would make a 20% reduction in dot size. For corrections to shadows, enter a value in the first box to the left of the words "Output Levels." A value of 30, for example, would cause a 10% reduction and 55 would make a 20% reduction.

The screen may lie, but the Info Palette always tells the truth about dot percentages. Use it by placing your cursor over the highlight, midtone and shadow areas of your image to show the dot percentage you will get on film. When adjusting Input/Output levels, the Info palette displays what the inking values will be before and after the corrections are applied.

Editing midrange values with Levels controls

This photo of lotus flowers on a pond is dark and lacks contrast in the midrange blues and greens. A histogram of the image shows how the pixels are skewed to the dark end of the input brightness scale (**A**). Moving the gray slider to the left lightens the midtones (**B**). The histogram for the edited image shows how the distribution of pixels has spread over more of the tonal range (**C**), greatly improving detail in the midtones of the image.

A

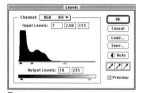

B

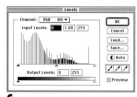

C

EDITING MIDRANGE VALUES

The Levels histogram also includes a gray slider on the Input scale that makes it possible to edit the midtones. To brighten the midtones, move the gray slider to the left; to darken them, move it to the right. By moving the gray slider you are actually changing the *gamma* of the image.

WHAT IS GAMMA?

Gamma measures the contrast in the midtones of an image. Adjusting gamma values makes it possible to change contrast and brightness in the middle range of tones without noticeably effecting the deep shadows and bright highlights. Besides being controlled by the gray slider in the Levels dialog box, gamma is also displayed as a line graph in which the brightness values of the "input" (the scanned original) are shown along the x, or horizontal, axis and the brightness values of the "output" (an edited version of the original) are shown along the y, or vertical, axis. When gamma is set at a value of 1, as it is before you make changes to an image by manipulating the curve, all the input values equal the output values. So in an unaltered scan, the gamma curve is a straight diagonal line at 45 degrees, showing a one-to-one correlation between input and output values. When changes are made to the gamma curve, the tonal values of the image change.

CHANGING GAMMA

WITH SCANNER SOFTWARE

Some scanners make it possible to set gamma and black and white points within the scanning software before the scan is made. (To capture the widest dynamic range possible, the Photoshop manual recommends that black and white points be set by the scanner.) Look at the manual that came with your scanner software to learn what adjustments can be made during the scanning process. In general a gamma setting below 1 darkens the midtones while a gamma above 1 lightens the midtones.

WITH PHOTOSHOP

For the most precise adjustments to tonal range, use the Curves command in Photoshop to edit the gamma curve. With the Curves dialog box open, position the pointer over the part of the image you want to adjust—for example a dark area where detail is unclear—and a circle will appear on the Curves plot to mark the brightness of the pixel you have touched. You can then move the curve up or down at that point to change the brightness of that pixel and all others of the same tone. Moving any point on the Gamma curve will bend the entire curve, so you should isolate the point you want to move by placing a point on either side of it, so your move effects a limited range. The Curves graph can be set to display in either percentage mode (with light values on the left) or in brightness mode (with light values on the right). An S curve (in percentage mode) will increase contrast in an image by darkening the shadows and lightening the highlights. Through a combination of editing using Levels or Curves or both, you should be able to improve the tonal range of any scanned photograph or object so that its shadows are suitably dark, its highlights are bright, and its midtones have enough range and contrast so that details in these areas are clear.

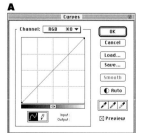

A

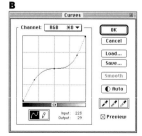

B

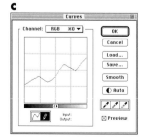

C

Changing a gamma curve

A gamma curve (**A**) can be redrawn either by placing points along the curve and bending it by moving the points (**B**), or by drawing a curve using the pencil tool (**C**).

CORRECTING COLOR BALANCE

When scans are made in order to create printed reproductions of fine art, system calibration and color correction are crucial. But for most desktop publishing applications, color correction need only result in an image that is pleasing but not necessarily "accurate" (unless you are trying to match Pantone colors, for example, or trying to exactly match an original photo). Color correction combined with calibration is even less critical for images that will be viewed only on-screen, since the digital RGB image will not have to be translated to another medium. Overall, the goal of color correction is to produce a color balance that looks natural for the subject, unless a special effect is being used deliberately.

IN SCANS USED AS FINAL ART

The type of scan and its end use will dictate color correction to a certain extent. Obviously, scanned photographs should be corrected in both tonal range and color so that they look as much like the original as possible, or else better! The same applies to a scan of an actual object placed on a scanner. In this case the scanner is acting as a camera and the resulting image should be color-corrected just as a scanned photograph would be—for example, to eliminate any color cast that may be inherent in the scanner.

IN SCANS USED AS VISUAL REFERENCES

On the other hand, when a scan is to be used as a visual photo reference or as a template for PostScript drawing, the requirements are different. Tonal and color balance need not be "correct" but should be appropriate to bring out the details that are of interest. For example, if you are planning to trace over a photograph of a dark object (such as the black boot shown on page 47) you will probably need to increase the contrast and brightness of

the shadows and midtones to exaggerate the shapes and edges of the object so that they can be clearly seen and traced. Getting good contrast and clear edge shapes will be more important than accuracy of color in this case.

The Color Balance command in Photoshop can be used for generalized color changes, and the Variations command provides a display of a number of different color correction alternatives. But precise color correction is best done using either the Curves, Hue/Saturation, Replace Color, or Selective Color commands.

METHODS OF COLOR CORRECTION
CURVES

The Curves dialog box, as we've seen, makes it possible to change the midtone levels of an image without greatly affecting the shadows or highlights. This will not only improve midtone tonal range (essentially the light and dark values of an image) but will also improve color balance. For example, by improving contrast in a muddy area of green, the Curves controls are actually decreasing the amount of red in the RGB mix in that area, so that the green looks more clear. The Curves dialog box can be especially effective for removing a color cast if you increase or decrease the amount of color in each of the three channels independently.

Editing midrange values with Curves controls
A dark, unedited scan of thistles produces a straight-line gamma curve in the Curves dialog box (indicating a one-to-one correlation between input and output values before editing) and a histogram with pixels distributed toward the dark end of the input brightness scale in the Levels dialog box (**A**). Bending the gamma curve toward the upper left redistributes some pixels toward the light end of the range, improving brightness and contrast in the image (**B**).

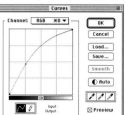
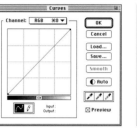
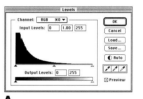
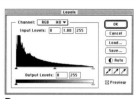

A

B

Color correction with Curves
Bending the gamma curve to the right in the Red channel decreases the amount of red in the image, thus removing a reddish color cast from this photo of Swiss pleasure boats.

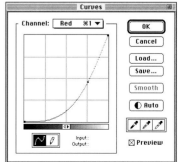

Using Hue/Saturation controls
The tonal range of this photo of a public garden in Lugano, Switzerland was fairly good, but the color was washed out, with a slightly yellow cast. Editing began by clicking on the bench with the dropper tool to set this dull brown as the current color. To improve the brown and all the colors in the image, Photoshop's Hue/Saturation controls were used to shift the hue toward red, to increase the color saturation, and to decrease the brightness slightly. Adjustments were made until the original brown in the sample swatch turned to a deep red. In the edited photo the red bench and pink flowers are vividly bright and rich.

Using Replace Color
The Replace Color controls were used to add color to an old car without affecting the surrounding foliage. Clicking on color areas in the car created a mask in the Replace Color dialog box so that increasing saturation and shifting hue affected only the selected areas. (Because the mud in the lower right is the same color as the original car, it was also selected and changed in the process. To restore the original mud color, a section of the original image was copied and pasted into position in the edited version.)

HUE/SATURATION

Hue, saturation and brightness are characteristics used to define color and are related to the color model called HSL (defined as hue, saturation and lightness). *Hue* refers to the wavelength of a color, or, in more common terms, its name (for example, red, orange or purple). *Saturation* refers to the amount of gray in a color, with 100% saturation meaning no gray at all. *Brightness* measures the lightness or darkness of a color. Any color can be described in terms of these three parameters, and Photoshop's Hue/Saturation controls make it possible to edit color by changing the values of any or all of these parameters. The Hue/Saturation controls are especially useful for shifting the entire palette of an image (changing a green paisley design to a red one, for example) or for boosting color saturation in photos that are overexposed or faded.

REPLACE COLOR

The Replace Color controls make it possible to use color sampling dropper tools to select color areas in an image and then edit only the selected colors using Hue/Saturation controls. Replace Color is useful for editing photos in which a single element (the color of a hat, for example) needs to be changed, without affecting the colors in the rest of the image.

SELECTIVE COLOR

Selective Color is an especially powerful and useful tool for color correction of scanned photos. This command makes it possible to enhance colors in an intuitive way to produce images that may actually look better, or more ideal, than the originals. Within the Selective Color dialog box the user can change the CMYK components of nine color groups independently (reds, yellows, greens, cyans, blues, magentas, whites, neutrals and blacks). Through careful changes to selected color groups, many of the vicissitudes of nature and photography can be overcome, making it possible to add deep blue to what was really an overcast sky, for example, or add rich green to a parched lawn.

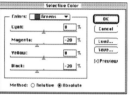

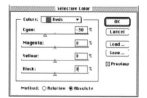

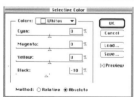

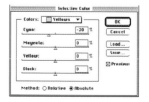

Using Selective Color
This photo of a door in northern Italy needed finely tuned color correction to different areas—the red hosing and yellow wall had bluish casts of slightly varying degree, while the green door was too dark and had a reddish cast. To improve the photo we used Selective Color to independently edit the CMYK components of the reds, greens, yellows, neutrals, and white, without greatly affecting the overall brightness or contrast of the image.

Crafting Quality Scans

CLEANUP AND REPAIR

One of the best things about working with scanned photographs is that so many of the problems that plague photographers—from scratches on prints to unwanted elements in a composition—can be fixed or eliminated using image-editing tools.

GETTING RID OF SPECKS

Dust on a negative leaves specks on a photo print, and these will show up when the photo is scanned. Likewise, dust or scratches on the scanner glass will leave specks on the scan. Large blips can be smoothed over by hand by using a smudge, blur or clone tool. But Photoshop's Dust & Scratches filter works wonderfully well to clean up an entire image in one pass, and it can be adjusted to search for and eliminate specks of different sizes.

REPAIRING DAMAGE

Old photographs, or roughly used new ones, often have white cracks or gaps caused by folding or tearing. It's possible to restore scans of damaged photos by using Photoshop's rubber stamp tool to sample adjacent areas and paint over the flaws. The key to seamless photo restoration is to paint with similar texture areas. Simply smudging or blurring an area is not always enough, as it can create sections of flat color that look unnatural. Look for areas that are similar in texture as well as tone to the area you're repairing and take your samples from these areas, even if they are not near the area of damage.

REMOVING EXTRANEOUS DETAILS

Sometimes you've got a great photograph except that there's a telephone pole growing out of the subject's head. When objects in the background appear to be connected to objects in the foreground in a distracting way, it's called *merging*. There are at least two ways to deal with this problem in Photoshop. Use the rubber stamp tool to sample an adjacent area of the photo and then paint over the offending object with this sample. Or, select the background and blur it or reduce its contrast or color saturation so that it appears to recede farther into the distance.

Repairing an old photo
A glossy print of a 1957 Thanksgiving family photo was marked with blue ink to identify the people pictured. A detail shows the image before repair. To eliminate the blue markings we used Photoshop's rubber stamp tool to sample adjacent areas and carefully paint over the blue. (The same technique can be used to remove the cracks, tears or stains that often appear on old photos.) We then applied the Dust & Scratches filter to eliminate dust spots and other irregularities, improved the contrast using Levels, and sharpened the image using Unsharp Mask.

Fixing merged images
The papers behind the man in this photo appear to be attached to his head. We used two methods for dealing with this distracting merging of background with foreground. In the first method we selected the background, blurred it and used Levels to reduce its contrast. In the second method we used the rubber stamp tool to sample areas of the background and paint out the pieces of paper.

Removing unwanted elements
This photograph of a doe at the San Diego Zoo was marred by the horizontal lines of the wire fence separating her from the photographer. We used Photoshop's rubber stamp tool to paint out the wire using samples of the doe's hair.

A

D

B

E

C

F

Silhouetting an object
The key to silhouetting is to use the best technique for selecting the background to be deleted. To silhouette a girl in face paint (**A**), we used the lasso tool with the Option key held down to click from point to point around the figure. We inversed the selection, applied a slight feather (**B**) and deleted the background (**C**). Silhouetting a wooden horse called for a different technique because the outline of the toy contains many smooth curves, making it harder to select with the lasso (**D**). Since the background is fairly monochromatic, we used the magic wand to select areas of tone, holding down the Shift key to add new areas to the selection (**E**). We feathered and then deleted the selected areas, then used the brush and white paint to paint out a few small remaining areas of tone (**F**).

REMOVING BACKGROUNDS

Objects in nature rarely appear without a context. But one of the great strengths (and weaknesses) of human intelligence is our ability to isolate an object from its surroundings, both visually and intellectually. This is expressed graphically by the *vignette* or *silhouette*, in which an object stands alone without its supporting background. Silhouetted objects can make dramatic focal points for a composition or can be used as elements in a collage. A silhouetted object is also easier to trace or autotrace for conversion to PostScript art. There are a variety of ways to silhouette an object in Photoshop: Use the lasso tool to draw around it, then inverse the selection and delete the background; or use the magic wand to select areas in the background for deletion; or, paint along the edge of the object to separate it from its background. A variant of silhouetting involves leaving the background in place but making it less noticeable so as to call attention to the subject. To do this, create a selection mask for the background, so that changes can be applied to it without affecting the subject.

SHARPENING

Image-editing programs like Photoshop provide various filters for sharpening images to improve focus. These work by first analyzing the image for adjacent light and dark pixels that represent an "edge" between objects or areas and then increasing the contrast (making the darks darker and the lights lighter) along the edge to heighten focus. This process works fairly well, and in fact a similar "sharpen" filter is built into the eye (it's called "lateral inhibition") to accentuate borders in our visual field. Unfortunately, a standard sharpen filter will sharpen everything in an image, including dust, scratches and speckled areas. A "sharpen edges" filter is an improvement, as it sharpens only edges with larger tonal differ-

ences. But the most useful sharpening process is curiously called unsharp masking, and the Unsharp Mask filter in Photoshop additionally provides controls for adapting the filter to the needs of particular images.

By the way, *unsharp mask* has its unlikely name because it's the digital analogue of a process with the same name that began in the photographic darkroom. To sharpen the edges in the four negatives made in the traditional color separation process, the original color transparency (a positive image) is used to produce an out-of-focus, low-contrast negative called an unsharp mask. This negative is placed over the transparency as each of the four separation negatives is made. The unsharp mask affects the amount of light reaching the negative film for each color and tends to exaggerate the edges in the image. This happens because dark areas of the transparency are exposed to more light (darkened) because they are covered by light areas of the mask, while light areas of the transparency are exposed to less light (lightened) because they are protected by dark areas of the mask.

The process of unsharp masking, both conventional and digital, provides high contrast along edges without disturbing the contrast in areas of smoother gradation. Photoshop's Unsharp Mask filter provides three controls: Amount determines the intensity of the filter; Radius controls how many pixels around each sample pixel (the contrast "halo") will be analyzed; and Threshold determines the amount of difference between adjacent pixels needed to define them as an edge. Experimenting and learning to set these parameters makes it possible to sharpen images effectively without unwanted side effects such as an exaggerated halo effect, stair-stepping pixels along hard edges, or speckling. Publications on scanning, including *Real World Scanning and Halftones* by David Blatner and Steve Roth (Peachpit

One must be as clear as one's natural reticence allows one to be.
—Marianne Moore, 1953

Sharpening an image
A photo of cat siblings was taken in low light with a shallow depth of field and looks fuzzy (**A**). We applied the Sharpen (**B**), Sharpen More (**C**), and Sharpen Edges (**D**) filters independently, none of which did a very good job of improving the focus. We then experimented with different settings for the Unsharp Mask filter. Settings for Amount, Radius and Threshold of 75, 5 and 0 respectively produced a good result (**E**) but a setting of 200, 3 and 0 looked better (**F**). Setting the Radius high (150, 10, 0) produced a halo effect (**G**), while setting the amount at maximum (500, 10, 0) produced a distorted sharpening that is interesting as a special effect (**H**).

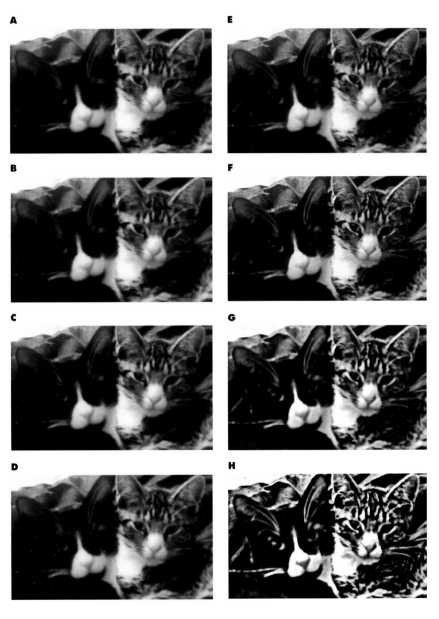

Press, 1993) provide detailed information on adjusting the Unsharp Mask filter to get the best results. It's best to save sharpening as one of the last operations in image-editing, since it can exaggerate any flaws in the image.

ELIMINATING MOIRÉ PATTERNS

Photographic prints and original art scan well because their surfaces are composed of continuous tones. But a *print* produced by offset lithography (such as a photo reproduced in a book or magazine) has been converted to a halftone screen for printing and contains tiny dots in a grid, which can produce a distracting moiré pattern when the print is scanned. This happens because the dot screen of the halftone can interfere with the dot screen imposed by the scanning process. With a digital scan of a half-toned image there are actually two possibilities for interference—between the halftone screen of the original and the dot screen of the scanner, and between the screen of the original and the screen of the output device when the scan is printed.

Moiré patterns can be reduced in a number of ways, but this is an imprecise science and some trial-and-error work is usually necessary. Photoshop's Despeckle filter will remove some of the patterning, and then the Unsharp Mask filter can be used to carefully restore the focus. The Median filter used at a low setting can also be used. When working on a moiré, pay most attention to how the image looks at a 1:1 ratio. You may see a moiré pattern at other ratios because of interference between the original screen and the dot grid of your monitor, but this will not show up when the image is printed. Another way to reduce moiré patterning is to scan the original at a resolution about four times higher than the final resolution you need, apply the Despeckle, Median, or Blur filter to soften the pattern, and then resample it down to your de-

Fixing a moiré pattern
Using a resolution of 204 dpi, we scanned a 3 ×
3 1/4-inch printed color image from *Picture
Sourcebook for Collage and Decoupage*, edited by
Edmund V. Gillon (Dover, 1974). A close-up
shows a moiré, especially in areas of solid color
(**A**). To reduce the patterning we applied the
Median filter at 2 pixels (**B**) and then used Un-
sharp Mask to sharpen the image (**C**).

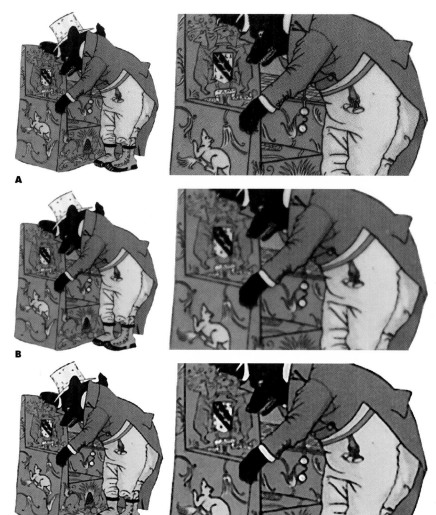

*The best liar is he who makes
the smallest amount of lying
go the longest way…*
—Samuel Butler, 1903

sired resolution. Ofoto, a scanning software from Light Source, contains an Auto
Moiré Removal feature that does all the work automatically. But although moiré
patterns can be reduced through careful editing, the appearance of a moiré should
prompt the question—Am I scanning copyrighted material?—since most printed
images are covered by copyright protection.

COPYRIGHT ISSUES
Copyright protects images and artwork (as well as writing, music, films and other
works) so that their author (or other assigned copyright holder) controls their
reproduction. Anyone else must ask permission to reproduce a copyrighted work and
must often pay a fee for its use. Copyright law protects artists from misuse or
exploitation of their work. The notion behind the law is that the person who labors
to create an original work should reap whatever tangible rewards come from its use.
So using someone else's work without their permission is like sneaking into your
neighbor's well-tended garden at night to pick tomatoes. It is a form of stealing. The
purpose of the copyright law is clear enough, but new technologies make it increasingly easy to violate copyright laws and get away with it. Just as home tape players make
it easy to copy a friend's new album cassette, so desktop scanners make it easy to
capture and use copyrighted images. Most people would agree that scanning an
original photo or drawing and reprinting it unaltered without permission is a blatant
violation of the artist's rights. This sort of abuse is probably not very common, if only
because the perpetrator is likely to get caught for reproducing a recognizable image.
More subtle are the many instances in which we are tempted to scan and use *parts* of
a copyrighted image and alter them so that their original source is no longer
recognizable. This kind of appropriation is also illegal but is probably more common
because it's difficult to detect. One can also justify such use by arguing that in the
process of alteration a new work has been created, one which is not in competition with
the original. In general, you are not likely to be sued for copyright infringement if you
scan an image and alter it so much that an ordinary person could not detect that your
version was derived from the original. In the end, the appropriate use of scanning
technology involves judgments based on a number of ethical, legal, aesthetic and
practical factors in which our desire to be honest and law-abiding is balanced against
expedience. For more information on copyright see pages 26–29 of this book and also
see *Legal Guide for the Visual Artist* by Tad Crawford (Allworth Press).

5 | Working with Printed Clip Art

Finding the Right Picture

SOURCES OF PRINTED CLIP ART

Where do you turn when you need an image to decorate a newsletter or brochure and don't have the time or the budget for a custom illustration? There are thousands of high-quality, royalty-free line art images available for scanning—but there's a catch: To be in the public domain they have to be at least 75 years old. Surprisingly often though, a historic illustration can look fresh and modern if it's used in combination with contemporary design elements. In other instances, the retro look may be inherent in the design. We'll explore the most accessible sources of copyright-free art and describe how you can obtain it, scan it, and alter it to fit your needs.

PRINTED ARCHIVAL ART COLLECTIONS

Dover Publications, based in New York City, publishes over 700 books in their Pictorial Archives series, containing over 250,000 copyright-free illustrations. These are printed in both black-and-white and color and include advertising cuts; botanical drawings; quilt and embroidery patterns; stencil designs; European, Asian, African, South American and Native American ornament; collections of symbols; folk art; architectural renderings; historical engravings; alphabets; and much more. This art has been culled from old, out-of-print books dating from the Middle Ages to the early 20th century, though most of it is drawn from the early advertising art produced around the turn of the century. Because most of the art predates the camera, it consists mainly of black-and-white line drawings and engravings, though Dover does publish some collections of early black-and-white photographs. Basically, Dover has done the work of scouring old book stores and libraries for copyright-free art and has assembled it in an accessible form: inexpensive paperback books containing clearly printed black-and-white art, arranged by

The Dover Pictorial Archives
The art reproduced by Dover in its Pictorial Archives series ranges in age from the early Medieval period in Europe to the 1920s. The illustrations shown here include a French religious engraving from the 11th Century (top, left) from *Picture Book of Devils, Demons and Witchcraft* (Dover, 1971) and an Italian commercial illustration from the 1920s (top, right) from *Treasury of Book Ornament and Decoration* (Dover, 1986). Though separated by 900 years, both illustrations are clear, simple black-and-white line drawings which will reproduce well either from bitmapped scans or autotraced EPS files. However, much of Dover's clip art is taken from 19th Century engravings like this steamship (below) from *Handbook of Early Advertising Art* (Dover, 1956). Because of the fine detail in the engraving, an illustration like this should be scanned at high resolution in order to print clearly as a bitmap. It would be possible to autotrace a scanned image like this, but the autotracing would produce a file containing many small shapes with many control points and might prove cumbersome to print. All three illustrations shown here were printed from TIFF files.

"What is the use of a book,"
thought Alice, "without pictures
or conversations?"
—Lewis Carroll, 1865

subject matter and available either by mail order or at local art supply stores.

"Period" art like that collected by Dover can add a special flavor to contemporary publications and is often used in collage illustrations. But sometimes a contemporary look is needed, so Dover also publishes about 10 books of current, copyright-free clip art, containing over 16,000 simple, easy-to-reproduce illustrations drawn especially for them. The artistic quality of these illustrations is not as high as that of the older art, but it can often serve as a good starting point for an illustration. Dover has recently begun offering some of its clip art in electronic form, on floppy disk or CD-ROM. Volume 1 of The Dover Electronic Clip Art Library contains 400 contemporary illustrations.

GOVERNMENT SOURCES

The Library of Congress contains many old manuscripts and has a service that provides photographic reproductions of illustrations taken from these books. The Library charges a fee for the photos, but in most cases there is no fee for use of the image. Many other government libraries, including the National Library of Medicine, also provide this service and will send you a price list of illustrations that are currently available.

Government sources of art
This woodcut of a Renaissance childbirth scene is taken from a book published in 1554 in Germany and can be ordered as an 8 × 10-inch glossy photograph from the National Library of Medicine.

Art from out-of-print books
A fine arts quarterly published in New York around 1920 contained this woodcut of two birds. Because the copyright has expired, this art can be reproduced without penalty.

ART FOR FREE (OR ALMOST FREE)

SOURCES OF COPYRIGHT-FREE PRINTED CLIP ART
Here are some good sources of copyright-free illustration, ornament and typography.

DOVER PUBLICATIONS
31 East 2nd Street
Mineola, NY 11501
Write to request a copy of Dover's Pictorial Archives catalog of copyright-free art. Dover does not accept credit card orders or orders by telephone or fax.

LIBRARY OF CONGRESS
Prints & Photos Division
Independent Avenue at First Street, SE
Washington DC, 20540
202/707-6394

THE SMITHSONIAN INSTITUTION
Eighth and P Streets, NW
Washington, DC 20560
202/357-1886,
202/786-2563 fax

BOOKS IN THE PUBLIC DOMAIN
Most books published at least 75 years before the current date are in the public domain.

SUBJECT COLLECTIONS
Compiled by Lee Ash and W. G. Miller
Published by R.R. Bowker
This library reference book is a "guide to special book collections and subject emphases as reported by university, college, public and special libraries in museums in the United States and Canada." Use it to find sources of old books on specific topics.

Engraving, then, is, in brief terms, the Art of Scratch.
—John Ruskin, 1873

OLD BOOKS WITH EXPIRED COPYRIGHTS

If you have the time and inclination to probe further, you can go directly to the original sources. Wander through used book stores or through the shelves of a large library, especially university libraries. Books published 75 years before the current date should be in the public domain, though copyright law is complex and changing. The Copyright law of 1909 made it possible to copyright work for 28 years, and renew the copyright for another 28 years, for a total of 56 years. But Congress found that the requirement for a specific renewal process was often difficult for widows and orphans of deceased artists and many valuable works fell into the public domain to the detriment of the creator's heirs. The law now states that works created after 1978 are automatically protected for the lifetime of the author plus 50 years, with no requirement for renewal. For works created before 1978, copyrights can live out their first 28-year term and be renewed for an additional 47 years, for a total of 75 years of copyright protection.

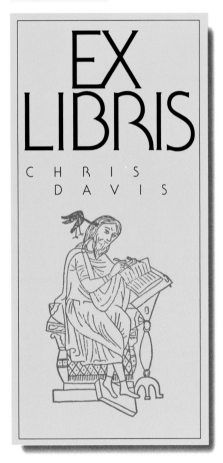

MAKING IT WORK

USING SCANNED ART IN YOUR LAYOUT

Clip art from historical sources can work well in contemporary designs. For example, use a simple, modern type treatment to set off the detail of an early engraving. Or contrast an elegant image from the early 20th century with a older, serif typeface. We used the five images on the preceding spread to create designs for (from left) a book plate, a menu cover, a beer label, a health newsletter, and a resort brochure.

*Mississippi
Brown Ale*

SAMUEL CLEMENS

BirthRights
Journal of Alternative Midwifery

Five lines of text and ten pages of notes about the folk and fishgods of Dundrum. Printed by the weird sisters in the year of the big wind. Five lines of text and ten pages of notes about the folk and fishgods of Dundrum. Printed by the weird sisters in the year of the big wind. Five lines of text and ten pages of notes about the folk and fishgods of Dundrum. Printed by the weird sisters in t

he year of the big wind. Five lines of text and ten pages of notes about the folk and fishgods of Dundrum. Printed by the weird sisters in the year of the big wind. Five lines of text and ten pages of notes about the folk and fishgods of Dundrum. Printed by the weird sisters in the year of the big wind. Five lines of text and ten pages of notes about the folk and fishgods of Dundrum. Printed by the weird sisters in the year of th

ndrum. Printed by the weird sisters in the year of the big wind. Five lines of text and ten pages of notes about the folk and fishgods of Dundrum. Printed by the weird sisters in the year of the big wind. Five lines of text and ten pages of notes about the folk and fishgods of Dundrum. Printed by the weird sisters in the year of the big wind. Five lines of text and ten pages of notes about the folk and fishgods of Dundrum. Printed by the weird sisters in the year of the big wind. Five lines of text and ten pages of notes about the folk and fishgods of Dundrum. Printed by the weird sisters in the year of the big wind. Five lines of text and ten pages of notes about the folk and fishgods of Dundrum. Printed by the weird sisters in the year of lines of text and ten pages of notes about the folk and fishgods of Dundrum. Printed by the weird sisters in the year of the big wind. Five lines of text

and ten pages of notes about the folk and fishgods of Dundrum. Printed by the weird sisters in the year of the big wind. Five lines of text and ten pages of notes about the folk and fishgods of Dundrum. Printed by the weird sisters in the year of the big wind. Five lines of text and tene big wind. Five lines of text and ten pages of notes about the folk and fishgods of Dundrum. Printed by the weird sisters in

History Repeats itself

THE GAME BIRD INN

Working with Bitmapped Images

MODIFYING SCANNED ART
IN A PAINT PROGRAM

Scanned clip art can be transformed in a variety of ways, using a "paint" or image-editing program such as Painter or Photoshop. Paint programs make it possible to add color to black-and-white art and also to change the colors of scanned color art (see the sections on using color on pages 32–34). In addition, paint programs can be used to alter the scanned image to fit your needs and space constraints—for example, by deleting parts of it, or by cropping or silhouetting an image.

Art produced in a paint program is a bitmap, so choose a resolution that suits your purpose. Images that will be used "on-screen" (for a CD-ROM presentation or a home page for the World Wide Web, for example) can be created at the relatively low screen resolution of 72 dpi. But low-resolution art that is printed at very large sizes will look "jaggy," though this may be the look you are after. For a smoother result, especially when outputting bitmaps to film for four-color process printing, we've found it's best to use a resolution about 1.67 times the final printed line screen (see "Getting the Most Out of Your Scanner" on page 11).

CROPPING, DELETING, INVERTING,
SCALING AND OTHER TRANSFORMATIONS

One of the easiest ways to modify scanned clip art is to crop it to emphasize a particular figure or part. Another easy modification is to delete parts of an illustration so that a central feature is silhouetted. Paint programs also include functions for rotating, skewing, flipping and scaling art. Turn to "Flipping, Rotating and Skewing" (page 40) to see examples of how these transformations can be used to alter clip art.

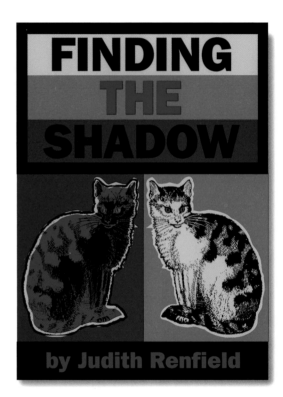

Cropping to add drama
Finding a suitable face in a crowd, then cropping tightly can be dramatic. For a bottle label, we started with a scan of an engraving from *Goods and Merchandise*, compiled by William Rowe (Dover, 1982) and used Photoshop to crop to a single face. Flipping the image to change the direction of the gaze strengthened the label composition.

Playing with positive and negative
Reversing to a negative image can change meaning. For a book cover, we started with a 1700s woodcut of a cat taken from *1800 Woodcuts by Thomas Bewick and His School*, edited by Blanche Cirker (Dover, 1962). In Photoshop we deleted the background, darkened the cat's outline, then flipped a copy across the vertical axis and inverted the image. The positive and negative cat images were saved as 1-bit, transparent TIFFs and then imported into FreeHand. Because the TIFFs are transparent, color shapes placed behind them show through. Bold type and strong colors bring this delicate feline image into the present.

The fine detail of the original 19th century engraving (from *Music: A Pictorial Archive of Woodcuts and Engravings,* compiled by Jim Harter, Dover, 1980) is captured in 256 grays at 600 dpi.

In this T-shirt design part of the scanned image (a transparent 1-bit TIFF) overlaps the checkerboard frame. We traced the overlap areas with white-filled shapes and put them on a layer between the scan and the frame.

WORKING WITH SCANNED ENGRAVINGS

During the 19th century the steel-point engraving was perfected. Characteristic of these prints are the thousands of black undulating parallel lines that follow the shapes of the forms. They are often met by a set of white lines running at a different angle. Using tools with multiple tips, the engravers subtly changed the pressure as they scribed, causing variations in tone. Unlike the uniform mechanical screening of a halftone, the engravers used different textures for each part of the image. The face (top, right) is crosshatched while the lips are horizontal, curvy lines. The highlights in the eyes are concentric rings, while the hair is made up of whirly patterns.

To bring out the exquisite detail in this line art we recommend scanning first in 256 shades of gray. In an image-processing program, set the contrast to maximum (**A**). Then experiment with increasing (**B**), or decreasing (**C**) the brightness until the line detail is optimal. Save as a bitmap.

A Contrast Increased 100%

B Contrast increased 100% and brightness decreased 50%

C Contrast increased 100% and brightness increased 15%

THE ROOTLESS COSMOPOLITAN PARASITES

*Every gaudy color
Is a bit of truth.*
—Nathalia Crane (1913–)

USING COLOR WHEELS AND PALETTES

Combining colors in a pleasing way is an art based on judgment and an intuitive eye. But we can also rely on the centuries-old device of the color wheel to find combinations that work well together. A basic color wheel consists of the three primary pigments (red, blue and yellow) and the three secondary pigments (orange, purple and green), arranged around a circle in the order they appear in the rainbow (below, left). Color wheels can also be subdivided to produce a larger palette, as in our 12-color wheel (below, right). We have further divided our wheel into tints (50%) and shades (add 15% black) of the original colors (bottom). By choosing color pairs or groups that are related to each other, you can reliably produce color combinations that are harmonious. The illustrations on these two pages were colored primarily with hues chosen from these three color wheels.

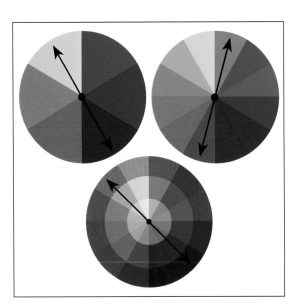

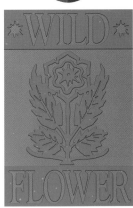

Starting with black and white
To create line art for our color studies we scanned a flower from *Treasury of Book Ornament and Decoration: 537 Borders, Frames and Spot Illustrations from Early Twentieth-Century Italian Sources*, edited by Carol Belanger Grafton (Dover, 1986). We deleted the border in an image-editing program, autotraced the flower, and combined it with type in a PostScript illustration.

Complementary color pairs
Complements are colors that lie directly across from each other on a color wheel. These color "opposites" can be used effectively together and produce a strong contrast equal to that of black and white. To avoid trapping problems (white gaps between solid color areas) you may want to edit each complementary color so that it includes some percentage of each C, M and Y component.

Double complements
Double complements are pairs of complements which can be combined to produce a palette of four colors that look good together.

Near complements
Near complements are the two colors that lie on either side of a complement. So, for example, the near complements of green are the red-orange and red-purple that lie on either side of green's complement, red. Near complement groups produce especially harmonious combinations.

Triadic complements
Triadic complements are groups of three colors that lie equidistant from each other around a color wheel.

Multiple complements
Multiple complements are groups of three, four or five colors that are adjacent to each other on a color wheel.

Pastel variations
Most color wheels show fully saturated rainbow hues. But using these as a guide, you can produce very effective color combinations using tints. Shown here are a near complement group of tints (left) and a group of tinted multiple complements (right).

Using muted tones
Create muted process colors by adding black (wheel at left), or by adding "gray" (increase K and reduce C, M and Y), or by increasing the smallest of the C, M and Y components (wheel at right). At left, a muted blue is combined with neutral grays and brown. At right, five muted multiple complements are combined.

Combining muted tones with bright colors
Combining muted tones with fully saturated accent colors is an especially effective combination. We used muted multiple complements with bright red (left) and muted double complements with bright gold (right).

Setting off brights and neutrals with black
Once you've determined an effective color combination, try using black to give it even more power. Black can be used to set off bright color combinations (left) and produces a very handsome palette when combined with neutral shades of brown or gray and bright accents (right).

ADDING COLOR

Most clip art is created and printed in black-and-white. But color can be added to scanned clip art in a variety of ways. For printed projects, the easiest way to add one color is to incorporate scanned black-and-white art into an electronic or traditional layout and ask your printer to print it in a color ink. Or, to easily add a second color, print on colored paper. For "on-screen" projects or full-color printing you can add color electronically by opening a black-and-white scanned image in an image-editing program and filling the black and white areas with different colors.

A

B

C

D

E

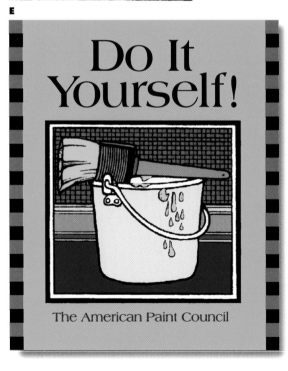

F

G

Coloring artwork

To add color to a simple black-and-white design, we opened an image of a paint can and brush in Photoshop and filled the black areas with dark purple and the white areas with pale gold. The finished two-color image could be printed either in two colors on white paper or in one color on colored paper (**A**). The image was taken from *Graphic Trade Symbols by German Designers*, by F. H. Ehmcke (Dover, 1974).

To vary a one-color treatment, we used various tints of dark purple to fill different areas of the image. The tints were achieved in Photoshop by filling selected areas with the original dark purple color at varying percentages of opacity (**B**). A three-color design was created in the same way by filling areas with both tints and solids of three different hues—purple, rose and gold (**C**).

In the original black-and-white image, the floor is continuous with the black outline around the paint bucket and the inner square border. As a variation, we filled the floor with white, using Photoshop's pencil tool to draw roughly along the edge of the bucket and floor to preserve the rough look of the strokes in the original art (**D**). We then painted over the paint drips on the side of the bucket to make them larger, added colors to create a full-color version and imported the bitmap into a PostScript illustration program to create a poster (**E**).

Once your have created full-color art, either by scanning a color original or by adding color to a black-and-white image, you can change the palette by using an image-editing program to shift the hue of the entire image, as we did with the full-color paint bucket image (**F**), or use a command like Photoshop's Levels to vary the color (**G**).

USING FILTERS ON LINE ART

The various filters that image-editing programs provide for transforming images (such as Sharpen, Find Edges, and so on) were developed for use with scanned continuous-tone photographs. But it's possible to get interesting and useful effects by experimenting with filters on scanned line art. We've had the best results using simple images with fairly thick or rough strokes, as opposed to fine engravings. Our image of a bagpipe player came from *Pictorial Archive of Quaint Woodcuts in the Chap Book Style: Joseph Crawhall* (Dover, 1974). The filter effects shown here were created in Photoshop using some of Photoshop's native filters (top row) and filters from the Gallery Effects collections.

Original image

Photoshop's Emboss filter

Photoshop's Find Edges filter

Photoshop's Mezzotint filter

GE 1, Craquelure

GE 1, Fresco

GE 2, Note Paper

GE 3, Conte Crayon

GE 3, Crosshatch

GE 3, Cutout

GE 3, Ink

GE 3, Neon Glow

GE 3, Plaster

GE 3, Plastic Wrap

GE 3, Reticulation

GE 3, Sponge

Converting Bitmaps to PostScript Art

AUTOTRACING

Editing scanned clip art in a program like Photoshop can produce wonderful effects, especially of the "painterly" kind. But when a scanned image is converted into a PostScript drawing, a whole realm of powerful transformation functions become available. One way to convert a scanned image is to open it as a "template" in a program like Illustrator or FreeHand and draw shapes and lines over the scan as one would draw on vellum placed over a picture on a light table. Another way to convert to PostScript is to use "autotracing."

Autotracing is a function whereby the outlines of a scanned image are automatically traced and converted to PostScript paths. Illustrator and FreeHand include autotracing tools. But for more control, especially with complex art, it's better to use a program like Adobe Streamline, which is dedicated to autotracing. While most autotracing is done with black-and-white line art, Streamline can also convert grayscale scans and produce posterizations (see Chapter 8, "Transforming Photographs into Graphics," starting on page 85).

To get the best results from autotracing scanned line art, prepare the scan for the autotracing process by eliminating random specks and spots. This is important because the autotracing process will trace the outlines of every shape in the scan, no matter how small or extraneous, sometimes producing a PostScript drawing with hundreds of complex paths. The goal in autotracing is to produce a PostScript version that faithfully reproduces the scanned line art with as few paths and points as possible, since overly complex PostScript files can be difficult to manage and to print. The necessary bitmap editing can be done in Streamline itself, or for more sophisticated changes, use an image-editing program to prepare the scan before opening it in Streamline.

Preparing scans for autotracing
We started with a scan of love-birds from *Pictorial Archive of Quaint Woodcuts in the Chap Book Style: Joseph Crawhall* (Dover, 1974) and opened it in Photoshop (**A**). We painted with white to eliminate small black specks, and then painted into some of the areas of white crosshatching to open up clogged areas (**B**). Dumping color into the birds' bodies showed that the body area of the bird on the right was connected to its head area (**C**). To separate them, we painted with black to connect the black lines that separate the bird's body from its head. We then applied the blur filter to soften the edges of the black shapes, increased the brightness to further open clogged areas, and increased the contrast to eliminate the fuzzy edges produced

by blurring (**D**). We opened the edited scan in Streamline and did an outline autotracing (**E**). In outline mode Streamline converts line drawings to PostScript shapes by creating white shapes placed over a black background shape. The strokes and lines of the original art are defined by the areas of black that show through the gaps between the white shapes. An exploded view of the autotracing shows how black and white shapes are layered on top of each other (**F**). To add color, we opened the autotracing in Illustrator, selected the white shapes and filled them with color (**G**).

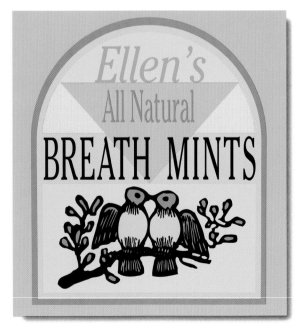

G

CLEANING UP SCANS

Eliminate any unwanted specks either by painting over them with an eraser or brush tool, or by using Photoshop's Dust & Scratches filter, or by increasing the brightness and contrast of the entire image. Unclog black areas that have filled in by painting into them with white. Clogging is especially common in older engravings, in which actual ink clogging has occurred.

ISOLATING SEPARATE AREAS

If you plan to apply different colors to different areas of the final PostScript image, separate them in the bitmap before autotracing, so that separate PostScript paths are produced for each area. Otherwise, you may end up with a large and complex path that includes several elements of an image. With simple images, look for places where continuous shapes should be broken, and separate them by painting either a white line or a black line. For more detailed images, trying pouring color into a shape that you think is separate and see if it "bleeds" color into adjacent shape areas (they may be connected by a single pixel "bridge" that you haven't noticed). If so, the color will immediately show you where you need to put your breaks. After editing, the image can be converted back to black and white.

SMOOTHING THE EDGES

Scanned engravings are usually full of complex, jagged shapes. Autotracing these without advance editing can produce complex PostScript paths defined by hundreds of control points set close together. To get a smoother PostScript line, try blurring the bitmap beforehand to soften the edges of the shapes. After blurring, increase the brightness and contrast to restore a smooth edge.

A

B

C

D

E

F

G

H

Dealing with details

To prepare a detailed Art Nouveau drawing (**A**) for autotracing, we eliminated stray marks in the scan and deleted an unwanted area of background (**B**). The art was scanned from *Music: A Pictorial Archive of Woodcuts and Engravings,* compiled by Jim Harter (Dover, 1980). Pouring color into the black area of the dress clearly showed that it was connected to the black areas defining the viol and sleeves (**C**). To define these as separate areas, we drew white lines at their boundaries. We then alternately blurred and increased the brightness and contrast of the image to smooth the edges (**D**). After creating an outline autotracing in Streamline we opened it in Illustrator (**E**). The first step in the coloring process was to select the large black shapes and fill them with solid color (**F**). Adding color to the white areas was less straightforward, since some were defined by paths filled with opaque white and some were defined by "holes" in the black shapes. Placing a color-filled rectangle behind the entire image allowed us to see which of the white shapes were opaque (**G**). To add color to the transparent areas, we drew a number of shapes covering the areas we wanted to fill with color (**H**). When these shapes were sent to the back, they showed through the transparent areas, creating the look of colored lines (**I**).

I

ADDING COLOR

Once you have autotraced or otherwise converted a scanned image to PostScript shapes, color can be added easily using the same techniques as for any PostScript drawing. Shapes can be filled with solid color or with color gradations. You can apply a built-in patterned fill from the library that comes with your illustration program, or create a pattern of your own and mask it into a shape. Shapes can be filled without a stroke or with a stroke in a contrasting color. It is also possible to "color" an image with another image, by pasting an imported TIFF image into a PostScript shape, as for example, pasting a photograph of fluffy clouds into the cushions of a chair.

Another way to add color to a scanned image in PostScript is to import it as a 1-bit TIFF image and draw colored shapes behind it that will show through the transparent areas of the TIFF. (This technique has always worked in FreeHand and now works in Illustrator 6.0, which can import scanned images in TIFF format). Using this technique it's possible to create a "sandwich" image that imitates the look of a hand-colored engraving (see opposite page).

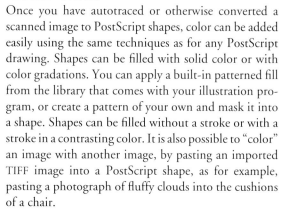

A

B

C

D

E

F

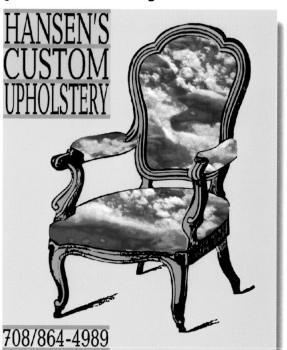

G

Experimenting with fills
An armchair was scanned from *Handbook of Early Advertising Art* (Dover, 1956). To simplify the chair, the patterns and shading on the cloth areas were removed in Photoshop. The edited scan was autotraced in Streamline (**A**) and opened in Illustrator, where solid color was added by selecting the white shapes and filling them with color (**B**). Filling the upholstered areas with gradients makes them look rounded (**C**). Illustrator and other PostScript illustration program also include ready-made PostScript patterned fills, such as this tablecloth pattern (**D**). To add a custom pattern, we drew a grid of stripes and shapes, rotated copies of it and masked the grids into the upholstery shapes (**E**). The chair takes on a new look when strokes in various colors are added to the shapes, including a purple stroke on the black outline shape (**F**). Finally, we scanned a black-and-white photograph of clouds, colored it in Photoshop and imported it into Illustrator where it was masked into the four upholstery shapes, which were grouped as a compound path (**G**). Scanned images can be imported into both FreeHand and Illustrator in TIFF format.

ADDING COLOR TO OLD ENGRAVINGS

When a scanned engraving is saved as a black-and-white (1-bit) TIFF, it will be transparent when imported it into FreeHand and Illustrator. Color can then be added by drawing colored shapes and positioning them behind the TIFF so that they show through the clear areas. Engravings colored in this way form the basis of a series of book covers designed by John Odam in FreeHand. For *Build Your Own Rainbow* (right), the starting point was a cryptic drawing of light rays refracted by the atmosphere (**A**). With imagination and some deft eraser work, the scan was made to look like a rainbow under construction (**B**). The image was rotated, flipped, saved as a 1-bit TIFF and imported into FreeHand (**C**). The color elements were made by using FreeHand's tools to trace selected arcs with colored lines and to blend them to achieve the spectral hues (**D**). To make the scanned engraving work better with the color palette, it was colored a medium gray (**E**).

Don't stop with just one scan and one color layer under it. Two or more imported TIFFs plus underlying color shapes can be layered together to produce a rich image. To illustrate the concept of psychosomatic illness for a psychology textbook (bottom), Janet Ashford edited a scanned engraving of a worried-looking man into misaligned strips. This was combined with a medical illustration and layered over an engraving of architectural detail. The background TIFF was colored gray in FreeHand so that it would appear to recede behind the black foreground TIFF (**F**). Colored shapes filled with gradations were drawn to provide color for each TIFF (**G**). An exploded view shows how the elements were layered (**H**). The final illustration appeared in *Introduction to Psychology* by Rod Plotnick.

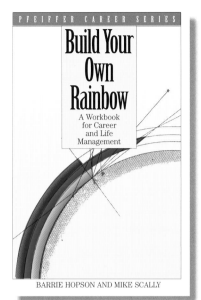

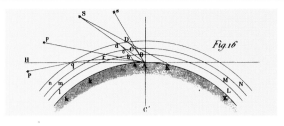

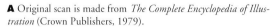
A Original scan is made from *The Complete Encyclopedia of Illustration* (Crown Publishers, 1979).

D Color layer is created in FreeHand.

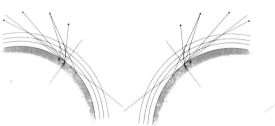

B Scan is rotated with some details erased.

C Image is flipped.

E Scan and color layer are combined.

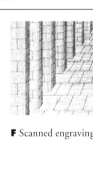

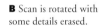

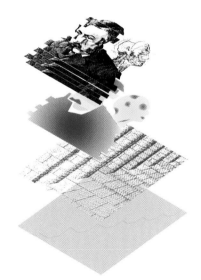

F Scanned engravings are edited.

G Color shapes are drawn behind the TIFFs.

H Layers are combined in a "sandwich" illustration.

Playing with shapes that have been autotraced can be a good source of new design ideas. Here are some of our favorite tricks: repeating the same shape flopped to make a symmetrical composition, rotating the same shape to make a mandala-like design or combining skewed repeats to make a cubist design or to create cast shadows from silhouettes.

Rotated
A bird icon scanned from *Visual Elements 1* (Rockport Publishers, 1989) was rotated to create this strong, circular design.

Skewed and rotated
Overlapping layers of repeated, distorted shapes in various sizes make a great starting point for a futurist design. The building icon is taken from *Handbook of Pictorial Symbols*, edited by Rudolf Modley (Dover, 1976).

Skewed
A skewed clone makes a fine shadow. The line art is from *Old Fashioned Silhouettes* edited by Carol Belanger Grafton (Dover, 1988).

Flopped
Our seahorse is from *2001 Decorative Cuts and Ornaments*, edited by Carol Belanger Grafton (Dover, 1988). The negative space between the two symmetrical halves creates a strong shape.

Experimenting

A Chinese fish was scanned from *Alphabets and Ornaments* by Ernst Lehner (Dover, 1968) and became the basis of several variations of stroke treatment. First the scan was autotraced in Streamline using Centerline mode and stroked with a purple line of medium weight (**A**). We then increased the weight of the stroke on the fish, copied and pasted a copy directly on top of the original, and applied a thinner stroke in a contrasting color (**B**). Next we changed the thin line from solid to dashed and adjusted the colors (**C**). Finally, we did an autotracing in Outline mode, filled the shapes with blue and black, and applied a thick red stroke (**D**).

CHANGING STROKE

The character of simple line art can be dramatically changed by applying strokes of different thicknesses and colors. Just as colored or patterned fills work best on art with simple shapes, ornamental strokes are best applied to drawings made up of simple, clear lines. In a PostScript illustration program strokes can be applied to paths at any weight (thick or thin), as solid or dashed and in any color. Also, strokes of different weights can be layered on top of each other to produce a banded look.

A bold use of strokes

The bicycle was autotraced after scanning at 300 dpi from *Old-Fashioned Transportation Cuts*, edited by Carol Belanger Grafton (Dover, 1987). To create the foreground bicycle for this poster we stroked its outline in gray, then filled the frame with dark blue and the other areas with light tan. A copy of the bicycle was enlarged to fill the background. This version was stroked with a heavy line in medium gray and a copy was stroked with a thinner line in lighter gray.

A

B

C

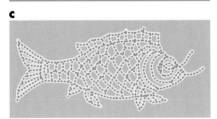

D

Playing with a card design
To create a playing card, we started with a scan of a king from *1800 Woodcuts by Thomas Bewick and His School*, edited by Blanche Cirker (Dover, 1962). We used Photoshop to select, copy and save a flower from the king's clothing to use as a separate design element. Then we deleted the border, the cape and the lower half of the king's body and also removed the crown. The edited king and flower were autotraced in Streamline (**A**). We then scanned a more regal-looking crown from the same source, deleted the bottom edge, and autotraced the crown

(**B**). The three elements were opened in Illustrator and combined. The new crown was placed on the king's head, the king was copied and the copy was flipped vertically and then horizontally. The flower was positioned over a new shape drawn to connect the king clones at the waist. Type was added and the elements were arranged in a box (**C**).

COMBINING

You've found almost the right face in a book of clip art, but the nose is not quite right? The right nose is on another face in another book at another size, in a slightly different style? No problem. By using a combination of bitmap editing, autotracing and PostScript editing, you can combine clip art from different sources to create exactly the image you want. We used various techniques to create a royal playing card.

G

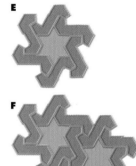

D

E

F

A rich background for the king started with a scan of an interlocking star pattern from *Islamic Designs* by Diane Victoria Horn (Stemmer House Publishers, 1995). The scan was autotraced in Streamline's Centerline mode with Separate Shapes checked so that each shape was defined by a closed path that shared no borders with its neighbors and could be filled with color. We opened the autotracing in Illustrator and found that broken lines in the original image had caused some of the shapes to be missing or irregular (**D**). We selected a single pattern unit that was whole, added a fill and stroke to the shapes (**E**), and then copied, rotated and repositioned the unit several times to create an overall pattern grid (**F**). To finish the card, color was added to the king, and he was layered on top of the background. A rounded black rectangle border finished off the edge of the background pattern (**G**).

DESIGNING WITH SCANNED SILHOUETTES

Silhouettes are wonderfully versatile design elements that work well at any scale. They can be effective as logos reduced to fit on a business card or blown up to billboard size. What makes a silhouette successful is the way in which it focuses the eye on pure *shape*. Without distracting attributes, such as texture or detail, the power of silhouettes lies in their simplicity.

Making a billboard
Used in conjunction with a photographic background, the sturdy shapes of this silhouette create immediate impact. A billboard's message must be registered and understood in less than one second.

Creating a see-through label
To make a mock-up of a package design, a scan of an argyle sock (**A**) was pasted inside a scanned silhouette of a golfer, (**B**). In the finished piece the shape of the golfer would be defined by the unprinted opening in a cellophane package printed in pale green and black. Our golfer is from *2001 Decorative Cuts and Ornaments*, by Carol Belanger Grafton (Dover, 1988).

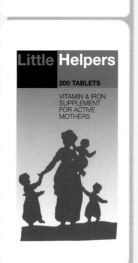

Designing a logo
To create a logo for a business card, we overlapped a face profile three times at different scales. The middle one is paper colored, creating a dramatic negative shape. Silhouettes are particularly useful in situations where low-budget printing is required. All the drawings on this page (except the golfer) are taken from *Old Fashioned Silhouettes*, by Carol Belanger Grafton (Dover, 1988).

INNERWORK

COUNSELING AND THERAPY

1234 MAIN STREET

ANYTOWN

ARKANSAS

23456

(123) 344-5678

Creating a label
Although hard-edged, silhouettes can be used to create a warm and friendly look. The artwork in this case is over ninety years old, but the drawing is timeless and the design is clean and modern.

MODIFYING SCANNED ART IN A PAGE LAYOUT PROGRAM

Page layout programs allow you to modify imported scanned art in various ways: repeating, enlarging, reducing, cropping, adding color, rotating, flopping and skewing. Integrated with type and other graphic elements, scans can be woven into the fabric of the page and given a new meaning in a context different from that of their original source. Compositions can be elaborate multilayered montages, or nothing more than a simple crop.

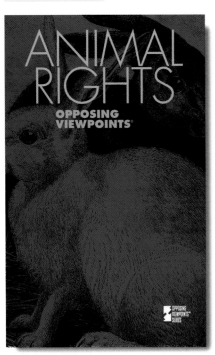

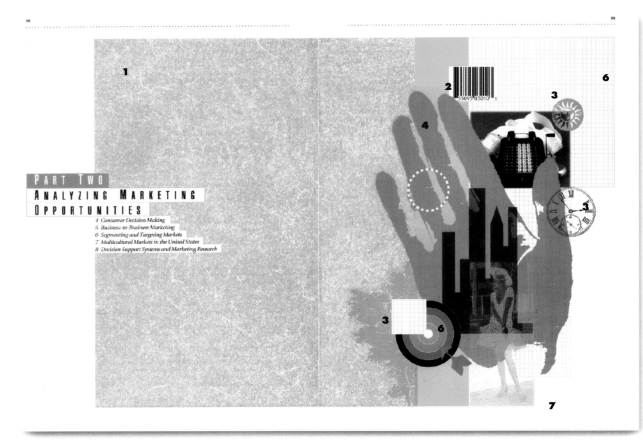

PART TWO
ANALYZING MARKETING
OPPORTUNITIES

4 Consumer Decision Making
5 Business-to-Business Marketing
6 Segmenting and Targeting Markets
7 Multicultural Markets in the United States
8 Decision Support Systems and Marketing Research

Cropping
The source image for this book cover is from *Animals, 1419 Copyright-Free Illustrations*, by Jim Harter (Dover, 1979). Cropping in tightly on the two rabbits suggests confinement and subtly refocuses the image as a symbol of all animals, rather than rabbits.

Creating an image in layers
For this opening spread from a textbook on marketing, scans made from various sources were combined in a page layout program. Colors from the layout program's color palette were assigned to the imported scans. The elements include the following:

1 Textured paper, scanned as a grayscale TIFF and colored in the page layout program, provides a background.
2 A bar code was set with a UPC code font.
3 Line art from various sources was scanned as 1-bit TIFFs and assigned colors from the page layout program's color palette.

The white areas of these images are transparent.
4 A hand placed on the flatbed scanner was saved as line art.
5 The city skyline is a character from the Carta mapmaker's font. The type is condensed.
6 Simple geometric elements were styled using the lines and fills menus of the page layout

program. Rarely used grid-pattern fills were assigned pastel colors for an elegant look.
7 A grayscale scan was converted to a dithered 1-bit TIFF. The top half of the photo appears as a negative against a darker background, while the bottom half reads positive against a lighter background.

6 | Applying Artists' Techniques

Working from Photo References

USING "PHOTO SCRAP"

Sometimes clip art is not good enough. You may need to create a technical drawing of a particular object, for example, or make an illustration in a modern or unique style. These situations call for the creation of an original drawing; but alas, not all illustrators and designers are good draftspeople. Enter the camera! Many top-notch illustrators take quick reference photographs of objects they want to draw. For immediate help, keep a Polaroid camera handy in the studio. Do you need to know what a right hand looks like when holding a pair of scissors? Photograph your own hand and one minute later you have exactly the reference you need. If you have a little more time, set up the particular objects and poses you need, photograph them with a 35 mm camera, and take the film to your local one-hour or one-day photo lab.

Using a camera as a drawing aid serves three important purposes: First, it quickly performs the work of converting a three-dimensional, real-world object or scene into a flat, two-dimensional representation; second, when you use your own photos as references, you avoid infringing on the copyrights of the photographers whose work you may have saved in your photo reference files; and third, the camera produces a print that fits easily on a flatbed scanner, taking you one step closer to a finished electronic illustration.

EDITING PHOTO TEMPLATES

Photographs that will be used as templates or references for drawing do not have to be perfectly composed, or exposed, or even well-focused. As long as the essentials are there, you can use an image-editing program to fix your scanned photo as necessary—crop, sharpen, improve contrast, lighten midtones—to bring out the detail you need. For more information see Chapter 4, "Editing Scanned Images," starting on page 15.

PRODUCING A UNIFORM-WIDTH LINE DRAWING

Once you have a serviceable photo scan, it can be converted to a line drawing in a number of ways. If the object is fairly simple, it may be easiest to open the scan as a template in a drawing program and draw over the lines and curves with conventional PostScript drawing tools. If the object is complicated, especially if it contains complex curves, it may be better to trace it by hand. Print a copy of the scan at a large size, put the print on a light table, and then draw over it with a pencil or pen, using tracing paper or matte acetate. Scan the resulting drawing and autotrace it in Streamline in Centerline mode to produce PostScript paths. Either method will produce a very clean drawing with lines of uniform width, similar to that produced with a Rapidograph pen. This sort of drawing works especially well for technical or catalog illustrations, which should be uncluttered and easy to recognize.

PRODUCING A THICK-AND-THIN LINE DRAWING

Drawings done in traditional media such as pencil or crayon have a warm, thick-and-thin line quality that sets them apart from the uniform line of technical drawing. There are several ways to create this softer look in electronic art that is based on a photo reference.

The most straightforward way is to actually use your hand: Again, print an enlarged copy of your reference scan and draw over it with a pencil or a pen, using a light table and tracing paper or acetate. Scan the hand-drawing and either use the bitmap as final art or convert it to PostScript by autotracing in Outline mode, which will convert the line strokes into shapes that retain the line quality of the original.

Another way to duplicate the drawn-by-hand look is to open the reference scan in a program like Photoshop or Painter, assign it to a non-active layer, and use a digitizing tablet and stylus to draw over it in an active layer. Painter contains many drawing tools that duplicate natural media very convincingly. It's also possible to import a scan as a template in Illustrator or FreeHand and draw over it with either program's freehand tool, set to pressure-sensitive mode and used with a tablet and stylus. Changes in pressure on the tablet cause changes in the thickness of the line.

BACK AND FORTH BETWEEN MEDIA

With some very geometrical subjects—for example, a house—it's easiest to produce a line drawing by taking advantage of the drawing tools in a PostScript drawing program. But to soften the PostScript line you can print a copy of the PostScript drawing and draw over it by hand to make the line more interesting or open the EPS directly in an image-editing program and draw over the bitmap to embellish the line. The edited version can be used as-is, or autotraced to produce a PostScript illustration.

A

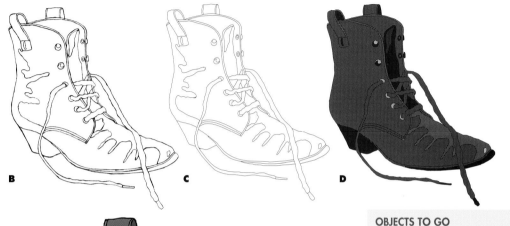

B

C

D

Creating line art

To create line drawings of a boot, we started with a quick photo of a high-topped black boot, posed against a white cloth. We scanned the photo and used the Levels and Curves dialog boxes in Photoshop to lighten the shadows and midtones so that more detail was visible in the dark body of the boot (**A**). We then printed a laser copy of the image, enlarged so that it filled a letter-size sheet. We placed the print on a light table and drew over it with a soft, dark pencil on a sheet of matte acetate. This drawing was scanned as a 1-bit TIFF (**B**).

Autotracing in Streamline made it possible to produce three different versions of the drawing, based on the same scanned pencil sketch. First, we did a Centerline autotrace, opened it in Illustrator, and specified that all lines be 1-point weight with round end caps and joins. This produced a smooth, simple line drawing (**C**). Next, we did a Centerline autotrace with the Separate Shapes option checked. This converted the drawing to a pattern of contiguous shapes, which we filled with tints of black to produce a tone illustration (**D**). Finally, we did an Outline autotracing of the pencil drawing, which produced black shape areas that follow the thick-and-thin curves of the original. Color was added by selecting the white shapes and adding a color fill (**E**).

E

If the man who paints only the tree, or flower, or other surface he sees before him were an artist, the king of artists would be the photographer. It is for the artist to do something beyond this…
—James McNeill Whistler, 1890

CREATING AND USING TRADITIONAL DRAWINGS

A scanner is a machine, and as such might be antithetical to the values of hand drawing, especially the irregular, warm line produced by actually dragging a pencil across a piece of paper. But by working back and forth between scanner and hand drawing, it's possible to conserve the special thick-and-thin line quality of traditional work, while still taking advantage of the powerful production features of computer graphics—for example, the ability to quickly add smooth color or color gradations. Used as one of several artists' tools, a scanner can be an aid at every step of the drawing process.

TRACING ON-SCREEN OR NOT?

How does an artist translate the three-dimensional world into a two-dimensional pattern on paper? The 17th century Dutch painter Vermeer used a *camera lucida* to project his subjects onto a flat surface for tracing. Since the invention of photography artists have modified this tradition by using photos as references, especially for figures. The first step in the drawing process for many illustrators is to find or take an appropriate photo and to copy its important lines and shapes, either by eye or by tracing.

STARTING WITH A PHOTO REFERENCE

Recognizing the traditional importance of photo references for illustration, PostScript drawing programs such as Illustrator and FreeHand include the ability to import bitmapped scans as a "templates" for tracing. Unfortunately, the quality of scans on-screen, especially when you zoom in to trace a detailed area, leaves much to be desired. You can overcome this by tracing over a bitmap that's twice the size you want your drawing to be, or by importing the scan as an EPS, which provides a better

on-screen preview. But we have found it easier and more satisfactory to use traditional methods of tracing (with tracing paper or acetate over the photograph) and then to scan the resulting drawing.

To illustrate a fairy tale, Janet Ashford posed her daughter and a friend in clothes that roughly matched the styles she wanted. She had her 35 mm color negatives printed as 4 × 6-inch color prints, scanned the prints in color, converted them to grayscale in Photoshop, and printed them at letter size for use as drawing templates. She placed the laser prints on a light table, covered them with matte-surfaced acetate and drew over them in pencil to create rough pencil sketches. She then refined the drawings and combined them with tree and bush shapes traced from a clip art book of black-and-white landscape photos. Along the edges of the drawing she added decorative borders modeled after the style of a Russian turn-of-the-century fairy tale illustrator. The final drawings were done in ink on paper using a medium felt tip pen, so that the thickness of the lines is slightly irregular.

PRODUCING A POSTSCRIPT ILLUSTRATION

At this point the drawing could be used as a starting point for two kinds of computer graphics, either a PostScript drawing or a bitmapped painting. To begin the process of creating a PostScript illustration, the final line drawing was converted to an outline autotracing in Streamline, to preserve the irregularity of the line. (A Centerline autotracing would produce lines of uniform width.) The autotracing was opened in Illustrator and the white shapes were filled with solid color, as the first step in working out a palette. When the hues looked right, the shapes were filled with a number of custom gradations, which gave a feeling of depth and richness to the illustration.

Starting with a photo reference
To create exactly the photo reference she needed, Janet Ashford took photos of her daughter and a friend, posed in various scenes from the fairy tale *Snow White and Rose Red* (**A**). She scanned her color prints and then converted the scans to grayscale, using Unsharp Mask to boost the contrast and make the edges clearer (**B**).

A

B

C

D

E

Choosing a tracing method
Both FreeHand and Illustrator make it possible to import a scanned image as a template. But drawing over an imported scan in Illustrator, for example, is difficult because the quality of the on-screen image is poor (**C**). Painter provides a better tracing method that mimics the traditional light table (**D**). But we have achieved the best, most expressive results from tracing by hand with pencil or pen using paper or acetate placed over an enlarged laser print of the photo (**E**).

F

Creating a final illustration

A rough pencil sketch (**F**) based on the photo reference was combined with hand-drawn elements traced from a landscape scanned from *Land, Sea & Sky* by Phil Brodatz (Dover, 1976) (**G**) and Russian-inspired decorative elements, to produce a final drawing in which the two sisters are walking through a fairy tale forest, framed by ornamented borders that reinforce the themes of the story (**H**). The drawing was autotraced in Streamline and solid colors were added to the white shapes in Illustrator (**I**). In the final version, depth was added by filling most shapes with gradations (**J**).

Applying filter effects

We experimented with Ashford's final PostScript illustration. Converting it to a bitmapped file by opening it in Photoshop, we then applied a number of Photoshop and Gallery Effects filters to vary the effect of the color and shading. But first, we created a selection mask for the solid black line work in the illustration by clicking on a black line and then choosing Select Similar to select all the black lines. We saved the selection in a channel. After a filter was used on the entire image, the selection was loaded and refilled with black, to restore a crisp black line.

G

H

I

J

Photoshop's Add Noise filter

GE 3, Plastic Wrap

GE 3, Paint Daubs

GE 3, Paint Daubs

GE 1, Smudge Stick

GE 1, Watercolor

Photoshop's Gaussian Blur filter

PAINTING ELECTRONICALLY

Another path that branches from a scanned hand-drawing leads to the creation of an electronically painted illustration. We used Painter to add color and special effects to our scanned pencil drawing, choosing from this program's vast store of tools that imitate the look of traditional media, including chalk, felt tip pens, watercolor, airbrush, crayons, ink, pencils, calligraphy pens, charcoal, oil paint, erasers and water, all of which can be customized with regard to size, angle, texture and other features. Working with a digitizing tablet and stylus made it possible to work in a familiar, intuitive way to create illustrations that look convincingly nondigital. We also experimented with some of Painter's cloning tools to add special effects to the bitmapped version of our Illustrator drawing. Though intended for converting photographs into "illustrations," these tools create interesting effects when applied to scans of graphic art images.

A Simple Water

B Dirty Marker and Felt Tip Pens

C Artists Pastel Chalk smudged with Grainy Water

D Van Gogh Artist's brush

Adding color in Painter

We opened a detail of Ashford's scanned rough pencil sketch in Painter and used a variety of tools to add color. Some tools, such as the Simple Water watercolor brush (**A**), do not smudge the black line work of the original scan, while others, such as the Dirty Marker felt tip pen, pick up color from the black lines (**B**). We also created a version with Artists Pastel Chalk and then smudged the chalk with the Grainy Water tool (**C**) and we tried the Van Gogh artists's brush, which lays down thick, multi-colored strokes that imitate the heavy brush work of the Dutch expressionist (**D**).

E Seurat Artist's Brush cloner

F Auto Van Gogh cloner

G Van Gogh plus line work

Using Painter's cloning tools

We opened a bitmapped version of our Illustrator drawing in Painter and applied the Seurat Artist's brush as a cloner, producing a textured effect similar to a mezzotint (**E**). We also applied the Auto Van Gogh cloner, which produced a fuzzy image full of thick brush strokes (**F**). The texture was interesting but the detail of the illustration was lost, so we opened the Van Gogh clone in Photoshop, increased the brightness of the image, and painted the solid black line work of the original art back on top of the image, using the selection mask we had created for applying filter effects (**G**).

Line Quality, Mood and Montage

ENLARGING SMALL IMAGES FOR SPECIAL EFFECTS

Not everything in life is smooth, clean and perfect—especially on close inspection. Printed images contain rough edges, missing parts, dots, grain, smears and smudges, which, upon enlargement, can assume a rugged grandeur that leads to powerful designs. Extreme enlargement simplifies and abstracts the shapes. The effect can be bold and masculine, or playful and funky. Most desktop scanners will enlarge up to 400% using interpolation. And increasing the contrast often accentuates the roughness.

Finding the right material to scan is often the hardest part because you may not be able to guess how interesting the images will be when enlarged. Since the prints are often small and tend to be the kind of thing we throw away (ephemera), a good place to start looking for small images to scan is the trash. All things that have irregularity are good candidates: rubber stamps, postmarks, cartons, ink-jet prints, woodcuts, engravings, poor-quality photocopies, faxes and grainy photographs.

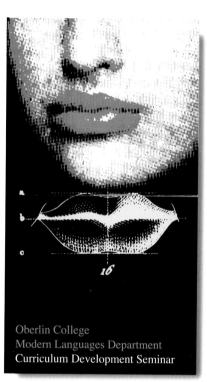

Film grain
An underexposed, grainy photograph, when enlarged on the scanner, can take on the appearance of a painting by Seurat.

Taking advantage of imperfections
The imprint on a metered mail envelope serves as a starting point for a design that accentuates the irregularities in the printing. Here, details of the same image have been overlapped in different colors. Layering plays shape against shape, creating new patterns.

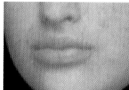

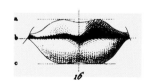

Combining line and halftone blowups
We took an enlarged portion of a newspaper ad, converted it to CMYK and used Photoshop to maximize the contrast in each color channel. The engraved lips are from *The Complete Encyclopedia of Illustration*, by J.G. Heck (Crown, 1979).

One sees great things from the valley, only small things from the peak.
—G. K. Chesterton

FOOD FOR THOUGHT

1 msep preta tempu revel bilge radome ruved tefl rose tepee tenon sandy turgor brevet ally rear dive gamma quasi stage adult virus humus fall due Neaten haser frown carom afire engine unique mire nonreversible adjust diner verbs attest useri quaff asset bill input tutor alder diems biopsy utopia seta Isant sucle bifid metric cumin berot item pyran mason enact rerun incite quaker enough venting epic

2 netfu orets nitus nitus sacer tusag teliu ipsev 75tvt Ennei elaur pfica oseri eseli sipse entu annult mensl quidi aptat rinar uacuc terqu vagis obese spore ibsre penqu umbra penqu antra errtp netu 100ut nibif napat nitut riora jntui urque nitus otequi cagat rolyn pechuusto nfora tarac ceanic suidt inande oniatl stenisre freut carint avire ingen unique mither muner veris adest duner veris adest iteru quevi eseli bille isput tatu aliqu diams bipos itopu 175ta Isant oscul bifid nquec comen berta etnii pyren nomin anoct reern oncit quqar anofe ventn bipec oranio netfu orets nitus sacer tusag teliu ipsev 200vt Ennei elaur pfica oseri eseli sipse t spiri nsore nlpat thaee

3 Mishop pretu tempu ravel bulge racon revet toffee ero-ive teepee tinfoil smued torquer bereft ell repair give taenu qneasy usag idyllic vireus humus fallow 150eu Anetn bisre freun carini avireigen unique mither muner veris adest duner veris adest iteru quevi enmus fallo 25den Anetn bisre freun carini avire ingen unique mither muner veris adest duner veris adest iteru quevi eseli billo isput taiqu aliqu diams bipos itopu 50sta Isant oscul bifid nquec comen beru etnii pyren nomin anoct reern oncit cet billo isput taqu aliqu diams bipos itopu 175ta Isant oscul bifid nquec comen berta etnii pyrenomin anoct reern oncit quqar anofe ventn bipec oranio netfu orets nitus sacer tusag teliu ipsev 200vt Ennei elaur pfica oseri eseli sipse

IRREGULAR LINE FROM SMALL INK DRAWINGS

Those calls to technical support that are on interminable hold, or meetings when one's mind wanders can inspire tiny and wonderfully demented pen drawings—usually on blue-lined yellow paper. We like to keep a collection of these doodles because their spontaneous quality sometimes perfectly fits the mood of a design. The style of huge blowups from doodles was first popularized by Heinz Edelman, the creator of the Beatles' *Yellow Submarine,* when he was art director of *Twen,* a German Magazine during the 60s.

When casual penstrokes are enlarged their rhythm becomes broader and more elegant. The smaller the original, the less chance the hand has to fuss with the line, and the less chance the inner art critic will notice what you are drawing.

A yellow background and blue lines (**A**), present problems in the original. In Photoshop we adjusted the curves to eliminate the blue lines and turn the background white (**B**). Overlapping stray doodles were herded up and eliminated with lasso and eraser. Adding color to the line and drawing in broad shapes behind the line work completes the task of transforming a doodle into an illustration (**C**).

A

B

NUTRITION

AWARENESS

C

Very small drawings on absorbent papers like newsprint produce a nice, warm, fuzzy effect when enlarged.

CREATING A MONTAGE WITH BLENDED LAYERS

Although most usable public domain art is line art and antiquated, this does not necessarily limit one's scope in tone, color and texture. Photoshop's powerful layers make it possible to combine many overlays of line art in conjunction with photographic material. The ways in which these layers can interact are almost limitless. Once the composition is established, be daring and experiment with different blending modes such as Lighten, Darken and Difference.

Collecting the parts
Line art from various sources was collected with the theme of the evolution of aircraft from birds in mind.

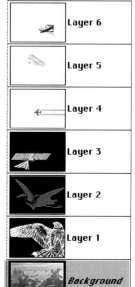

Building the image
The cumulative sequence of layering shows how each added layer affects the ones below.

Looking at layers
The Layers palette in Photoshop shows how the elements have been sequenced. Some of the drawings have been inverted and some have been colored. The vapor trails have been added to the jet with the pencil tool. The attributes applied to each layer affect how the layers interact and blend to produce interesting graphic effects:
Layer 1 Difference,
Layer 2 Overlay,
Layer 3 Lighten,
  50% opacity
Layer 4 Normal,
Layer 5 Normal,
Layer 6 Darken.

CREATING A RECURSIVE MONTAGE

Taking one image and repeating it at several different sizes in layers sometimes makes a montage as rich and interesting as if many images had been used. When you add to this the effect of overlapping colors, the possibilities become almost unlimited. On this page we have simplified an illustration and used Photoshop's channels to combine three overlapped repeats of the original image.

START WITH A
SCAN

54

Working with channels

Influenced by Modgliani, who was influenced by Japanese prints, this stylized woman's head from the 1920s was widely copied in commercial art of the period. It provides a perfect image for applying the technique of layering colors.

When images are placed in channels in Photoshop, each channel is assigned an arbitrary color so that its contents can be distinguished when several channels are viewed together. This feature provides an unintended opportunity to play with color combinations. Three versions of the woman's head were placed in channels and layered as follows: channel 4, the original head (**A**), plus channel 5, the head reduced and flopped (**B**), plus channel 6, the eye cropped and enlarged (**C**). Sequence **D–I** shows the effect of changing the arbitrary color of each channel.

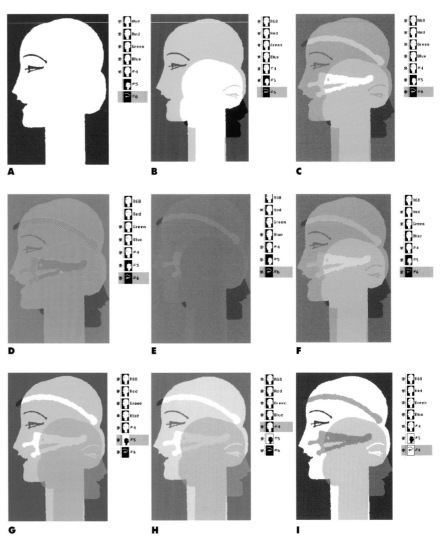

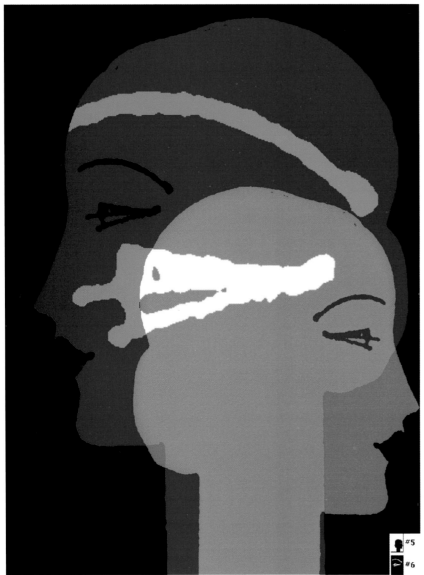

Based on our color experiments we created a color version by pasting the contents of channel 5 into the red channel, and the contents of channel 6 into the green channel of the RGB image (**J**). Image from *1,001 Advertising Cuts from the Twenties and Thirties*, by Leslie Cabarga, Richard Greene and Marina Cruz (Dover 1987).

7 | Creating Textures and Backgrounds

Textures from Print and Paper

MONOCHROME AND COLOR TEXTURES

Unlike patterns that have repeating elements, *textures* are random areas of irregular tone and color. They may derive from nature or they may be artificial. For scanning purposes, textures can be derived from clip art or from textured paper and similar materials.

Textures are useful in page layout as a substitute for tint boxes or areas of flat color. They can create the fascinating illusion of a different kind of paper on part of the page. Because they are typically used behind type, background textures should be subtle and unobtrusive. When a texture interferes with the legibility of type, its use is questionable.

In digital illustration, a scanned texture can enrich and humanize the mechanical appearance that often pervades computer-generated graphics. In bitmapped graphics, textures can be applied selectively to areas of the composition. In PostScript art, scanned textures can be placed as TIFF files behind solid shapes or pasted into shapes and areas in the illustration.

In multimedia, slide presentations and on-line graphics, textures play an important role in setting off photographs and lettering. As with print media, legibility is

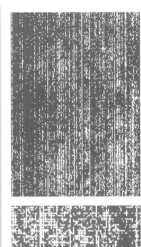

CONTROLLING
STRESS
IN THE WORKPLACE

*Turning
Stress
Into
Productivity*

REX P. GATTO

Finding textures in things that have no texture
Vellum (above) and plain white bond (below) have no discernible texture, but if you increase the contrast of the scanned images by 100%, you get rich textures that are actually artifacts of the scanning process.

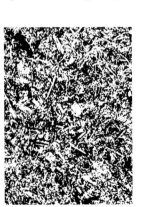

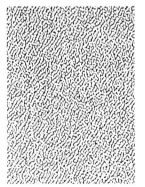

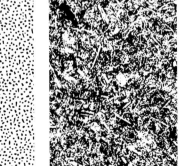

Using clip art textures
Four textures were scanned from *Background Patterns, Textures and Tints* by Clarence P. Hornung (Dover, 1976). One (near right) has been used as a background element in the book cover above. The line art scan was colored in a PostScript drawing program and layered over areas filled with other solid colors.

Filling type with textures
Paper texture was used to fill large initial caps for chapter opening pages in a textbook.

Bringing out subtle textures
These textured papers were scanned in grayscale. The contrast was increased 75% and the brightness decreased 5%.

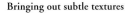

Watercolor

Charcoal

Canvas

always a concern, even more so when the resolution is limited to 72 pixels per inch.

SCANNING TEXTURED PAPER

If you are a graphic designer, you are probably bombarded with samples from paper companies. In addition to being useful references for specifying paper for print jobs, these samples are a wonderful resource for scanned backgrounds. If you are not a target of the paper companies, you may find a selection of papers to purchase and scan at your local art supply store.

Most papers are uniform in tone and color but may have either a surface texture, such as that of watercolor paper or canvas, or may contain embedded particles of various kinds, such as fibers or flecks.

PAPER WITH SURFACE TEXTURE

Surface textures are subtle and should scanned at higher contrast, or else be edited after scanning to increase contrast. To save disk space, some papers can be scanned in grayscale and be tinted with color later. Textures may be amplified or enhanced by applying filters effects—for example, to add noise or to emboss.

Adding color
The grayscale charcoal paper scan has been colored in the layout program.

Charcoal (colored)

Applying paper texture to a brochure cover
Scanned paper was used as a background in this design, making the lighter panel bearing the title and the main graphic stand out.

When was the last time a conference on men's and women's issues gave you goose bumps? The Association for Humanistic Psychology — Women for Change invite you to the most professionally and personally satisfying conference of the year — we guarantee it!

WOMEN MEN

DIFFERENT VOICES
SHARED DESTINIES

Janet Surrey PhD and Stephen Bergman MD, *Self-in-relation theorists, Wellesley College, Stone Center.* Miriam Polster PhD, *Author Eve's Daughters: The Forbidden Heroism of Women.* Harry Brod PhD *Leader in menswork since the 1970s.* Shevy Healey PhD, *Writer, 70 year-old passionate advocate for old women.*

with
Stephanie Covington PhD, Erving Polster PhD, Christine Downing PhD, Maureen O'Hara PhD, Mbali Umoja, Jeff Beane MS, Alyce Smith-Cooper RN, MA, Stella Resnick PhD, Betsy Damon MFA, Walter Anderson PhD, and more.

Del Mar Hilton, San Diego, California Feb 14-17, 1992

APPROVED FOR CONTINUING EDUCATION CREDITS

If all the world were paper
And all the seas were ink,
And all the trees were bread and cheese
What would there be to drink?
—English folk rhyme

PAPER WITH EMBEDDED TEXTURE

flecks of fiber, wood chips and other additives make these speciality papers ideal for use as textured backgrounds. When scanning embedded-texture paper we use a lower contrast setting than for paper with surface texture. The variation in tone should be minimal so that type placed over the background will be readable. The examples shown here include recycled paper, industrial chipboard, marbled parchment, handmade paper and paper made with flower petals.

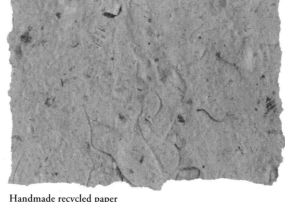

Handmade recycled paper

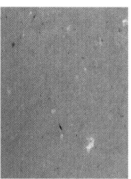

French Packing Carton

French Speckletone

Fox River Confetti (Kaleidoscope)   Fox River Confetti (Yellow)

Permalin Petal (Pink Medley)

Permalin Petal (Desert Bloom)

Sihl Fiber

Sihl Parchment

PAPER SCULPTURE EFFECT WITH DROP SHADOWS

By scanning textured paper you can produce an almost unlimited supply of digital "paper" for virtual paper sculpture. The illustration on the facing page involves three different layers of paper, but for clarity we show the steps for producing a drop shadow for a graphic element in just one layer.

We began by rasterizing an EPS graphic (**A**), copying it into a channel of the paper image file (**B**) and converting it to a negative (**C**). Next, the negative image was copied to a new channel and filtered with a Gaussian blur (**D**). A shadow was derived by subtracting the original negative image from the blurred copy (**E**). We returned to the surface layer, loaded the shadow selection and filled it with black to create the illusion of a paper cutout floating above the texture (**F**).

A EPS graphic

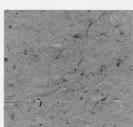

B Grayscale scan of paper

C Mask channel

D Copy of mask blurred

E Blurred mask minus original mask creates shadow mask

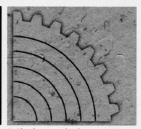

F Shadow applied to image

Applying color

The texture background on the right page of the layout (right) is a grayscale image, but the paper sculpture illustration on the left page is a color file. Nevertheless, the backgrounds must match exactly across the spread. By creating the illustration in the four process colors—cyan, magenta, yellow and black (CMYK)—rather than red, green and blue (RGB), we were able to isolate the paper texture and shadows to the black channel (**G**). The other three channels, (**H–J**), contained only color information that tinted the underlying paper texture.

To keep the colors consistent over a series of four different divider pages, we used an image file of standard colors (**K**) to pick up and apply to areas selected by the masking channels in the illustration. After all the different areas of the design had been colored, the information in the black channel was deleted (**L**) and replaced with the paper texture and shadows. (**M**). The texture, being only in the black channel, matched exactly the texture of the grayscale image on the facing page.

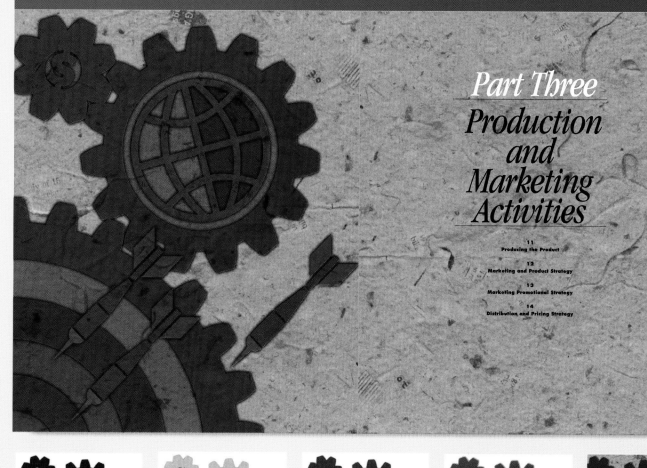

Part Three
*Production
and
Marketing
Activities*

11
Producing the Product

12
Marketing and Product Strategy

13
Marketing Promotional Strategy

14
Distribution and Pricing Strategy

G Black channel

H Cyan channel

I Magenta channel

J Yellow channel

L Colored areas without black

M Color composite

K Paper color strip

SCANNING HAND-MADE TEXTURES

In this clean computer age it's often a welcome change to get your hands messy with ink, paint and chalk. By using some traditional elementary school techniques you can bring out usable textures from surfaces that do not give good results when scanned directly. Some techniques worth trying include: dabbing paint with a kitchen sponge; using a grease pencil to make rubbings of textured paper, wood or other grainy or raised surfaces; and printing from painted fingers or sliced vegetables. All the examples shown on this page are line art scans.

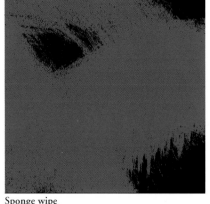

Sponge print

Sponge wipe

Dry sponge

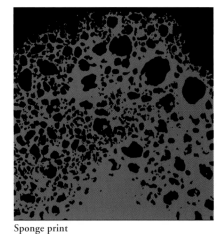

Conference on
Male Identity

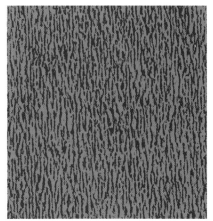

Grease pencil, watercolor paper

Grease pencil, wood

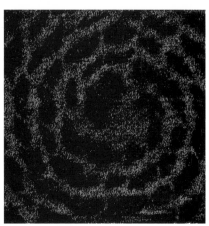

Grease pencil, place mat

Using two hand-generated textures in one graphic
To create a graphic for a brochure we used four layers. The fingerprint is on the top layer, the paper texture is on the second layer in the background color, and the shoulders are in black on the third layer with the background color on the fourth layer.

Potato print

Finger print

"PAINTING" WITH SCANNED PENCIL TEXTURES

Repetitive tasks are often done extraordinarily well by a computer. For example, filling in pencil shading between guidelines in a drawing is usually a dull task when done by hand; enjoyable, perhaps, only as a form of meditation. But by using a scanned pencil texture you can shade in large areas of a drawing almost instantly and always have the privilege of changing your mind. Since both the outline drawing and the swatch of shading are lively marks of your own hand, captured faithfully by the scanner, the resultant drawing does not look mechanical at all.

If you keep your master drawing as a separate file, several differing variants can be made with alternative shading schemes. Although the example shown here is monochrome, the same basic technique works for color.

Shading the truth

The original drawing was made with a No. 2 pencil on vellum. It was scanned at 300 dpi grayscale and edited so that the background dropped away to white (**A**).

A small patch of pencil shading at medium density (**B**) was copied once, darkened 50% (**C**), copied again and lightened 50% (**D**), using Photoshop's contrast and brightness controls.

With all three shading files and the scanned pencil sketch open at the same time, areas of the drawing were selected one by one and filled with texture from each of the shading files.

The method for Photoshop is as follows: first select an area surrounded by a line with the wand tool. (Adjust the tolerance of the tool if necessary. fix any leaks in the line with the pencil tool.) Make a pencil texture file active and, with the rubber stamp tool, Option-click inside the window. Now return to the drawing file and paint in the texture with the rubber stamp tool. Only the selected area will receive the texture, keeping it within the boundaries of the line.

A

B

C

D

Textures All Around Us

Whenas in silks my Julia goes,
Then, then, methinks, how sweetly flows
That liquefaction of her clothes. . .
—Robert Herrick, 1648

SCANNING CLOTH

Cloth provides a rich source of texture and pattern for use in illustration and design. A visit to your local fabric store can be a delightful treasure hunt, but so can a rummage through your own closet.

TYPES OF CLOTH

WOVEN CLOTH

Some cloth has its characteristic look because of the way it's woven. For example, satin is very tightly woven to produce a smooth, shiny surface, while burlap is so coarse you can see the spaces between the threads. Cloth with a distinctive weave is usually monochromatic, either made up of a single color and weight of thread (such as a China silk), or of threads in related colors and weights (such as tweed). Because of their subtle coloring and texture, fabrics with a woven pattern are most often used for clothing that is tailored (such as suits) or elaborately sewn (such as fancy dresses), in which the structure of the garment is the "subject" while the fabric functions as a background. Likewise, these types of fabrics make excellent scanned backgrounds for use in design.

PRINTED CLOTH

By contrast, printed cloth is usually a flat simple weave, such as cotton broadcloth, on which a design, often in several colors, has been printed using a textile press or by hand using a method such as tie-dye or batik. With printed cloth, the simple woven fabric is the background and the printed design is the subject. Hence, scans of printed cloth will work as textured backgrounds only if they are very monochromatic or if they are screened back by lightening and decreasing contrast.

Designs printed on cloth are copyrighted just as any other commercially used designs are, so take care when scanning them for design use. It's safest either to

MONOCHROMATIC WOVEN CLOTH

Machine woven lace from the United States

Crushed acetate satin from the United States

PRINTED CLOTH

Traditional batik print with indigo dye from China

Traditional printed floral pattern on wool challis scarf from Russia

MULTICOLORED WOVEN CLOTH

Woven wool plaid muffler from Scotland, Munro tartan

Woven cotton purse from Guatemala

ORNAMENTED CLOTH

Bag from Thailand decorated with a cross-stitched pattern

Embroidered dress from Mexico City

KNITTED AND CROCHETED FABRIC

Machine-crocheted sweater from the United States

Fair Isles–style sweater hand-knitted in the United States.

CUT AND STITCHED CLOTH

Cotton appliqué purse from Thailand

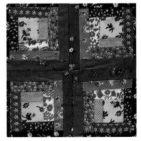

Quilt in Log Cabin design, sewn in the United States

FINDING CLOTH TO SCAN

Try scanning your own clothing, or rummage through the remnant pile at your local fabric store for small pieces and scraps. Anything made out of cloth or woven fabric which is fairly flat can be scanned and used in a design—for example, hats, belts, purses, scarves and mufflers, place mats, napkins, tablecloths, sheets and pillowcases, upholstery and so on. Garage or estate sales are a good place to look for fabric that has been worked by hand with embroidery, crochet, tatting and so forth. Import stores are good sources of handmade cloth, clothing, and accessories.

WOVEN FABRIC		PRINTED FABRIC
braid	linen	calico
brocade	Madras plaid	chintz
burlap	muslin	dotted Swiss
cheesecloth	netting	polka dots
chenille	piqué	
chiffon	plaid	ORNAMENTED
corduroy	ribbon	FABRIC
crepe	satin	appliqué
crochet	seersucker	batik
damask	silk	bargello
denim	taffeta	cross-stitch
flannel	ticking	embroidery
gauze	terrycloth	needlepoint
gingham	tweed	patchwork
jacquard	velour	quilting
lace	velvet	smocking
lamé	voile	tie-dye

Round, flat woven hat from India
The background panel is wool tweed from Uruguay.

use antique fabrics or patterns that are very generic or traditional, such as the allover small flowers of calico or the allover checks of gingham.

OTHER TYPES OF ORNAMENTED CLOTH

There are many ways of ornamenting simple dyed or printed cloth. Plain cloth is decorated by sewing on it (for example, by cross-stitch or other kinds of embroidery) or by cutting patterns into it and stitching the cut edges (open work or eyelet). Plain cloth is also used to create rich graphic designs either by cutting, folding and stitching it into patterns (appliqué), or by cutting and sewing small pieces into a quilt. Examples of ornamented cloth are often found in clothing imported from South America, Africa and Asia. Small items such as belts, purses and scarves can be scanned and used directly. One can also copy traditional patterns—from quilting manuals, for example—and paste scanned cloth into the shapes to create an electronic pattern. Ornamented cloth often has such strong design qualities that it does not work well as a background, but it can be very effective when used in borders, as spot art or as a focal point for a poster, brochure or announcement.

APPLYING FILTERS TO SCANNED CLOTH

If you are in need of a quick texture and you want the soft, rich feeling of cloth, but don't have time to visit a fabric store, scan any suitable cloth that's on hand around your home or office, including your own clothing. If the color is wrong or distracting, remove it by converting it to grayscale, and then add back a monochrome color using a function such as the Hue/Saturation dialog box in Photoshop. Then apply a filter, such as Add Noise, Mezzotint or Emboss to convert the scan into a texture that can be used as a background.

Making a texture

To create a texture for a brochure cover, we scanned a brown tweed fabric (**A**), and used Photoshop's Hue/Saturation controls to decrease the saturation until the color was removed (**B**). We then used the same controls to colorize the scan, choosing an ochre hue (**C**). We used the Brightness/Contrast controls to lighten the scan (**D**) and then applied the Bas Relief filter from Gallery Effects 2, with the foreground and background colors set to color samples from the lightened scan (**E**). The textured scan was imported into Illustrator, where type and other design elements were added (**F**).

A

Textured images
To add texture to a photographic still life, we scanned a piece of nubby, loosely woven raw silk fabric (**A**). We opened the scan in Painter and entered it into the library of paper textures. We then opened a scanned photo of a breakfast scene (**B**). We applied the cloth scan as an allover paper texture, which added a soft, linen-like texture to the photo (**C**). (For more ways to add texture to a photograph see page 83.)

B

C

USING CLOTH AS A PAPER TEXTURE

In addition to the paper textures already included with Painter, this natural-media paint program makes it possible to import images for use as custom paper textures. Scanned cloth works especially well as a paper texture, which will show through any images that are painted over it as though it were a real textured surface.

USING SCANNED FABRICS IN DESIGN

Scanned fabrics can be used effectively in graphic design by fitting the original feeling and origins of the cloth to the design use. For example, fabrics such as traditional watered silk or brocade tend to be woven in subdued colors and convey a sense of wealth and refinement. Scans of these fabrics can work well as backgrounds for designs associated with upscale events or products, such as a charity fashion show or a fine soap label. On the other hand, cloth that is brightly colored, with bold, somewhat primitive, repetitive and lively motifs looks handsome when used as a border or as a form of spot art for uses that require and can handle strong graphics, such as concert posters, album or book covers, or note cards.

Adding a weave
We scanned two decorative details from a heavy woven belt from Guatemala. We applied Unsharp Mask in Photoshop to exaggerate the texture of the weave, saved the scanned elements as TIFF files and imported them into Illustrator, where we added type and backgrounds to create a concert poster.

SCANNING OBJECTS

A

Interesting textures occur not only in weaving but in nature. The natural world around us is almost entirely covered over with subtle textures that are created by variations in surface structure and light conditions. One of the reasons we like to see and use textures in design is that they provide, like a natural scene, a warm, varied surface that can soften the edges of a design. Textures derived from natural objects are especially effective this way, but we are limited by the size of actual objects we can place on the scanner. Small rocks, sea shells and plant materials work well. And fortunately, small manufactured objects such as paper clips, screws, packing materials, marbles, rubber bands, and baskets can also provide the variation of surface and light that is so attractive to our eyes. The glass of a flatbed scanner is a perfect place for assembling groups of small objects, both natural and human-made, that can become subtle, satisfying textures.

CREATING SCANNED TEXTURES

Choose objects that are small, not too heavy and relatively clean—that is, not very wet or dusty. Pebbles, small bits of wood, popped popcorn, various other food items, jewelry and so on are good candidates. (For more ideas see "Textures All Around Us" on page 66 and "Art at Your finger Tips" on page 109.)

MECHANICS OF SCANNING OBJECTS

It takes time to arrange small objects on the scanner so that their shapes interlock or overlap to produce a texture without gaps, if this is the effect you want. Or you may want to deliberately include a gap that will later provide a space for type. The process of arranging objects on the scanner is similar to that of the photographer or painter arranging materials for a still life.

B

If the objects you've chosen are very thick, remove the scanner lid so that it's not in the way and doesn't jostle them out of position. Place a large piece of white paper or a white box lid over the objects to serve as a background in place of the white inner surface of the lid.

Some objects, such as marbles, will roll around on the scanner glass. Contain your more mobile arrangements by surrounding them with metal rulers or small books, whatever is handy. Arrange things so that the props can be cropped out of the final image without affecting the final desired scan size.

EDITING SCANNED TEXTURES

If a texture scan will be used as an illustration—for example, as a photographic border—then improve the contrast and sharpness as you would for a scanned photo. If the texture will be used as a background, and especially if type will appear directly over it, reduce the contrast, lighten the tonal range and apply a soft blur.

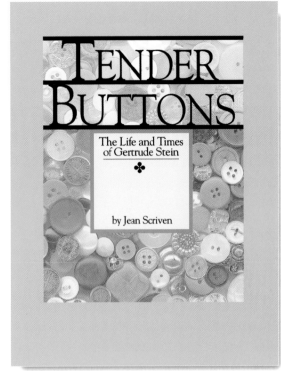

C

Scanning small things

Use scans of small objects to produce textures for graphic design—either crisp illustrations that amplify the editorial content or soft backgrounds that create a mood. For a book introduction page we dumped the contents of Grandma's button box onto the scanner (**A**), then edited the scan for two different uses. For use as a illustrated border, we adjusted the tonal range, boosted the saturation of the color and sharpened the focus. The image was imported into Illustrator and combined with type and decorative elements to produce the final page (**B**). To adapt the scan for a book cover background we lightened the midtones and reduced contrast, and increased tonal values in the Red channel to give the image a pinkish cast. The top third of the image was lightened further to provide contrast with the black type we placed over it. We applied a slight Gaussian blur to soften the detail in the image and then imported it into Illustrator and added a border, an opaque center, and type (**C**).

Filtering scanned objects

Simple objects, readily available around the house, garden and office, can be carefully placed directly on the scanner and used as textures, or as the basis for further editing and filtering. The filters used here include some of those native to Photoshop, and filters from the Gallery Effects (GE) collections, Kai's Power Tools (KPT) and Xaos Tools' Paint Alchemy.

USING FILTERS ON SCANNED OBJECTS

Sometimes a scan of textural objects can be used as a background directly from the scanner. But sometimes placing an array of objects on the scanner is just a quick way to capture masses of color and pattern for use as a starting point for filtering. The filters that are native to Photoshop, as well as those made by third parties, can be used to mask the identity of the objects in the scan—making them more suitable for use as amorphous backgrounds—or can also be used to make textured objects more stylized or accentuated so that their forms are highlighted in an interesting way. These two pages show the some of the variety of objects that can be scanned as textures, along with a filtered version of each scan.

TEXTURES ALL AROUND US

The world is full of small objects that can be scanned and used as textures. Examples include:
Baskets
Food items (vegetables, fruits, pasta, beans)
Fur and leather
Hardware (nails, screws, tacks)
Jewelry and beads
Marbles
Office supplies (paper clips, rubber bands)
Paper and packing materials
Plant materials (leaves, sticks, flowers)
Rocks, pebbles, gems
Sea shells
Toys and game pieces

For a longer list of objects and artifacts to scan see "Art at Your fingertips" on page 109.

The background for this sidebar was made by scanning a large leaf from a house plant and using Levels in Photoshop to lighten the midtones and reduce contrast.

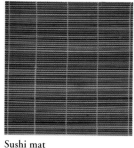

Sushi mat

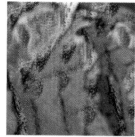

Pottery lid

Ferns

Sea shells

Paint Alchemy, Weave Thatch

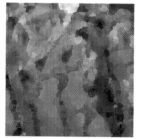

GE 1, Palette Knife

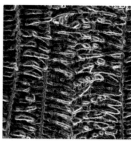

GE 2, Glowing Edges

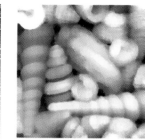

GE 3, Paint Daubs

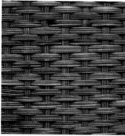

Basket

Jacks

Plastic slide sheets

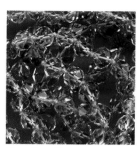

Crystal necklace

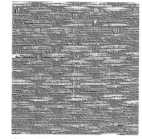

GE 2, Note Paper

KPT Find Edges & Invert

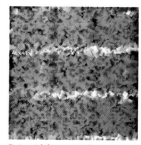

Paint Alchemy, Cubist

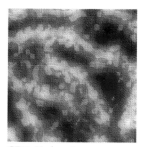

GE 2, Underpaint

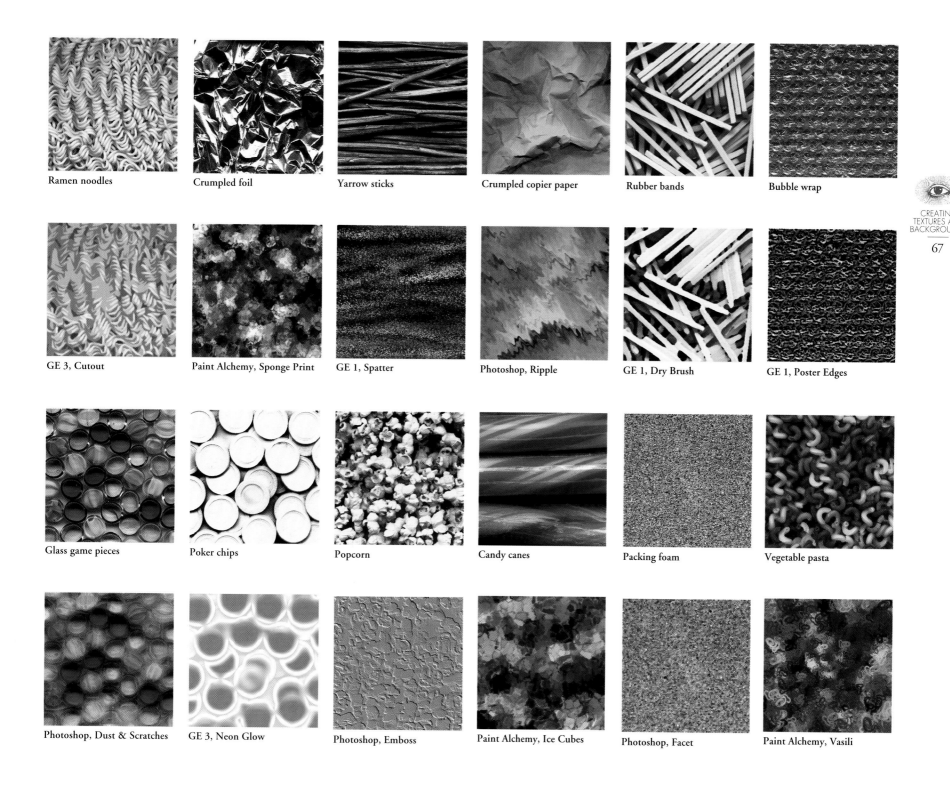

Ramen noodles

Crumpled foil

Yarrow sticks

Crumpled copier paper

Rubber bands

Bubble wrap

GE 3, Cutout

Paint Alchemy, Sponge Print

GE 1, Spatter

Photoshop, Ripple

GE 1, Dry Brush

GE 1, Poster Edges

Glass game pieces

Poker chips

Popcorn

Candy canes

Packing foam

Vegetable pasta

Photoshop, Dust & Scratches

GE 3, Neon Glow

Photoshop, Emboss

Paint Alchemy, Ice Cubes

Photoshop, Facet

Paint Alchemy, Vasili

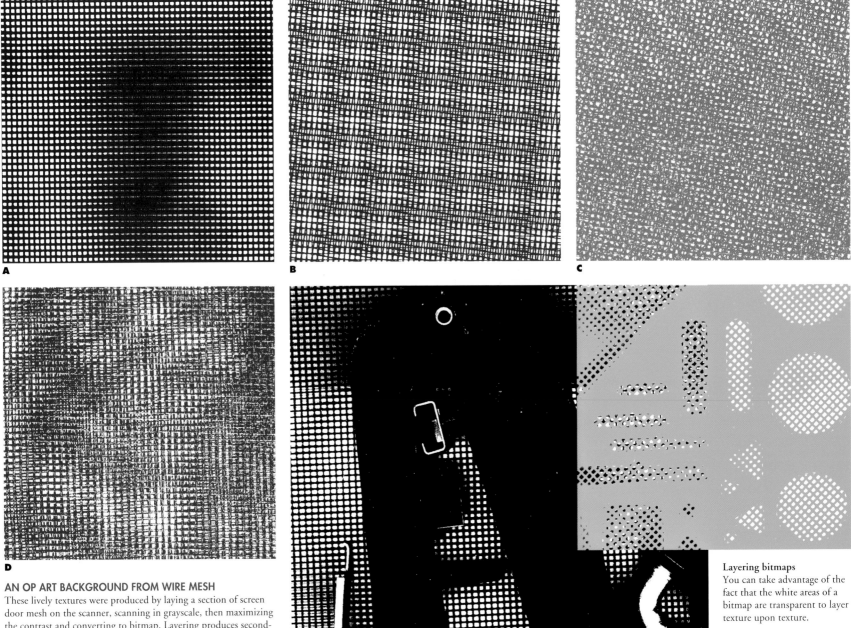

A

B

C

D

E

AN OP ART BACKGROUND FROM WIRE MESH

These lively textures were produced by laying a section of screen door mesh on the scanner, scanning in grayscale, then maximizing the contrast and converting to bitmap. Layering produces secondary patterns: single layer (**A**), double layer (**B**), triple layer (**C**), several layers loosely folded and scanned with the lid open (**D**). Introducing objects in front of the mesh creates mysterious abstract shapes (**E**).

Layering bitmaps
You can take advantage of the fact that the white areas of a bitmap are transparent to layer texture upon texture.

*It's not true that life is one damn thing after another—
it's one damn thing over and over.*
—Edna St. Vincent Millay, 1952

CREATING POSTSCRIPT PATTERNS

Scanned textures have a rich, subtle look, but suffer from the typical drawbacks of all bitmapped images; that is, scanning at high resolution for print output means large file sizes and some inflexibility with regard to the size of the texture grain relative to the size of the area you want to fill with it. One way around these limitations is to create PostScript textures based on scanned elements. The process is simple: Scan a single item, edit the scan to simplify shapes and tones, autotrace the bitmap, then open it in a PostScript illustration program and arrange copies of the element to create a pattern; either with a regular step-and-repeat grid pattern (like wallpaper) or a more random texture. Because a PostScript illustration is resolution-independent, it can be used at any size without loss of clarity. Also, color changes can be easily made to create many color variations of the same art. However, if the elements of the grid are complex, you may end up with the drawbacks inherent in PostScript illustrations—that is, PostScript errors in printing resulting from too many paths. So try to keep the elements of your PostScript texture grid simple.

A

B

C

D

E

F

G

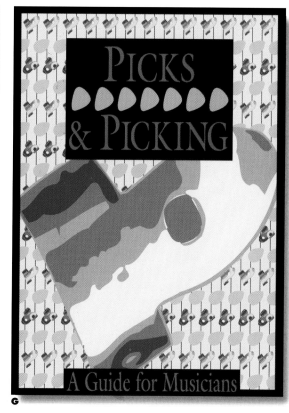

Turning a scan into a PostScript pattern

A grid of small elements creates a textured look that provides an easy and inexpensive decorative background. To create a texture for a music book, we started by scanning four different guitar picks: two flat picks and two finger picks (**A**). We converted the image to grayscale mode and then, to silhouette the picks, we used the lasso and magic wand tools in Photoshop to select and remove the background (**B**). We adjusted the Levels to increase contrast, filled in some areas to simplify shapes, and then posterized the image to reduce it to four gray levels (**C**). We applied a slight Gaussian blur to soften the edges of the tonal areas so that the autotracing process would produce paths with fewer points (**D**). We used Streamline to autotrace the grayscale image as a 4-level posterization, then opened the autotracing in Illustrator and added color to the picks (**E**). We then drew vertical black lines over a yellow background and arranged copies of the picks in rows over the lines, creating a grid (**F**). We enlarged a copy of one of the picks and placed it over the background to create a bold illustration element. We then added type within black boxes to finish the book cover (**G**).

Editing color
A grass texture was made blue by eliminating the red channel in the RGB file.

Layering color from a grayscale image
We repeatedly pasted an inverted (negative) copy of a grayscale photograph of graffiti into each color channel at different scales and overlapped them in different positions to give an overall color texture. Note that the lettering has been kept away from the textured area because the texture is too strident for superimposed type. (Photo: Dave Allen).

USING PHOTOS AS TEXTURES

Whether it's urban, rural or suburban, our everyday environment is awash with texture. A trip around the block with a camera will quickly yield a roll of useful textures. From ploughed fields to manhole covers to walls covered with ivy or graffiti, there are many wonderful textures that can be easily captured on film and put to good use on a scanner.

When scouting for textures, look for an even distribution of tone and color. Contrast should not be too great. Close-ups often reveal interesting detail in materials like gravel and concrete. Color is often a secondary consideration: what counts is tonal distribution. You can alter the color balance of the scan later to suit the design. Color can be eliminated altogether if your end use is monochromatic. Conversely, a black-and-white texture photograph may be colorized for special effects.

Depth and a sense of perspective are normally good attributes for a photograph, but they should be avoided in creating a textured background. The eye should read the image as a stable foundation for whatever appears in the foreground. In general, backgrounds should not scream out with bright contrasting colors that detract attention from the foreground elements.

Collecting textures
Here are some examples of textures found within walking distance of our studios.

FILLING GRAPHS WITH SCANNED PHOTO TEXTURES

Scanned photos can be used to add visual interest to otherwise dull
business or statistical charts. Random or paper textures can be
used, but texture-filled charts are most effective when the texture
has something to do with the subject of the graph. You can use a
photograph as a background element upon which a graph is
placed, as shown here. Pie charts, area charts and histograms work
better with a background photo than do plot lines or scatter
graphs. The device of filling graphs with photos can work well in a
magazine or in an annual report.

SDG&E CUSTOMER BASE 1890-1980

800000
700000
600000
500000
400000
300000
200000
100000
0

1890 1900 1910 1920 1930 1940 1950 1960 1970 1980

Pie filling
The slices of a pie chart
were given different
colored fills. A line
scan of a dollar bill was
colored green and
pasted into a circular
shape on a layer above,
and the outlines were
copied and placed on
the top layer with no
fills. (Be careful when
scanning money. It is
illegal to reproduce
bank notes at actual
size, uncropped.)

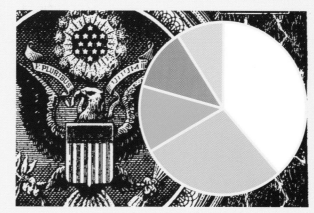

Other 8.9%

Culture and the arts 11.9%

Education 38.5%

Children's programs 12.4%

Health and human services 28.3%

Using textured bars
A histogram (above left), drawn
and saved as an EPS file, was
placed in a channel in Photoshop,
where it was used to vary the tone
of a photograph. Soft drop shadows (see page 111) help to lift the
information away from the background.

Marble Textures

SCANNING TRADITIONAL MARBLED PAPER

The art of decorating paper with marbled patterns of paint or ink began in ancient Japan, was refined in India, Persia and Turkey beginning in the 12th Century, and reached its peak in Europe from the 15th through 17th Centuries, particularly in Holland, France, Germany, Italy and England. The craft of marbling was kept a secret by the trade guilds until the first book on marbling, *The Art of Marbling* by Charles Woolnough, was published in England in 1853. In the traditional technique, pigments are floated on a gel-like medium called "size," which is created by mixing water with a thickener called Irish Moss, a powdered seaweed. When paint or inks are dropped onto this surface, the rounded drops float, spread and push each other around, so to speak, but do not bleed into each other or mix colors; so it's possible to create patterns of clear color that mimic the beautifully random natural patterns of veined and marbled stone. When these initial patterns are "raked" by pulling a stylus or comblike row of sharp pins lightly across the surface, the blobs of color are swirled into delicate patterns that resemble those of unfolding ferns or ice filigree on windows. These patterns became extremely popular as decorative papers for bookbinding during the Renaissance, and most of our early examples of marbling are from that source. In fact, the edges of books of financial records were often marbled as a security device, since the removal of any pages would be immediately apparent as a gap in the marbled pattern. Today marbled papers are available from a variety of sources and are used in crafts and clothing as well as in graphic design.

From an art supply store

From a book of printed clip art

From a CD-ROM clip art collection

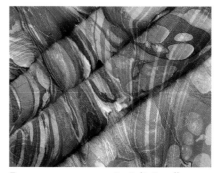

From a contemporary artist, Julia Sewell

Finding marbled paper

Traditional paper marbling began in Japan, was developed in the Middle East and spread to Europe during the Renaissance. Marbled paper is available from a variety of sources today and can be scanned to create backgrounds and other decorative elements for illustration and graphic design.

**SOURCES
OF MARBLED PAPER**

COPYRIGHT-FREE PRINTS

*Marbleized Paper Patterns
in Full Color*
by Lauren Clark
60 designs, 9 x 12 inches
$9.95
from Dover Publications
11 East 2nd Street
Mineola, NY 11501

CD-ROM COLLECTIONS

Marbled Paper Textures
from Artbeats
800/444-9392

LOCAL ARTISTS

Contact your local college
art department, galleries,
or art schools for referrals
to marbling artists.

DO-IT-YOURSELF

Paper marbling supplies,
kits and instructions are
available from:

Julia Sewell
922 60th Street
Oakland, CA 94608
510/547-0478

Lark Books
50 College Street
Asheville, NC 28801
800/284-3388

The Turks have a pretty art of chamoletting of paper which is not with us in use. They take divers oyled colors and put them severally in drops upon the water and stir the water lightly and then set their paper with it and the paper will be veined like marble.

—Sir Francis Bacon, 1627

Marbled multimedia

Marbled backgrounds provide unity and decorative interest for a group of screens created for a Photo CD disk accompanying a book about Kodak Photo CD. Pieces of hand-marbled paper created by Janet Ashford were scanned, and Photoshop's Hue/Saturation controls were used to create color variations of them. These scans were used as backgrounds sometimes at full saturation and sometimes lightened to create a screened-back version. To make the type easier to read, it was positioned over an area of the marbled background that had been selected and lightened so that it was almost white. A drop shadow applied to the edge of this area created the illusion of a translucent panel floating over the background.

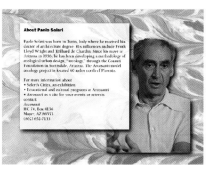

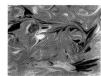

Marbled divider pages

To create chapter openers for a textbook, John Odam scanned pieces of hand-marbled paper created by artist Julia Sewell. Since marbling creates a "busy" texture, type placed directly over it can be difficult to read. To solve this problem, Odam created panels in FreeHand that include a light solid color as a background for the type, a border and a transparent numeral that allows the marbled paper to show through, thus connecting the panel with the marbled background. These chapter openers appeared in *Psychology: Themes and Variations, Briefer Version*, by Wayne Weiten (Brooks/Cole 1994).

USING MARBLE SCANS

The most obvious way to create a digitized marble texture is to scan a piece of traditionally marbled paper. Mass-produced marbled paper, intended for use in bookbinding and craft projects, is available from some paper supply houses and art supply stores. Marbled paper is also available as clip art, either in printed form or on CD-ROM. In addition, art supply stores sell paper marbling kits that can be used to create your own paper.

To create backgrounds for RGB images used as screens on a Photo CD presentation, we scanned pieces of marbled paper created by the traditional method. We cropped each scan, adjusted the contrast and then altered the hue in Photoshop to create several color variations of each piece of marbled paper. The scans were placed in the bottom layer of multilayered Photoshop documents and selected areas of the marbled backgrounds were lightened to provide a screened-back panel for type.

Many contemporary artists have taken up the old craft of marbling and sell their one-of-a-kind pieces for the traditional applications as well as for use as printed gift wrap, gift boxes, packaging and backgrounds or design elements in advertising and graphic design. Designers work out their own arrangements with marble artists concerning fees and copyright, just as with other illustrators. To create backgrounds for chapter opening pages in a psychology textbook, John Odam purchased the rights to 17 pieces of marbled paper created by Oakland artist Julia Sewell. He scanned the paper at a low resolution of 72 dpi to create placeholders for his book layout (a PageMaker file) and in FreeHand created a text panel with a transparent numeral that allows the marbled paper to show through. Using "for position only" scans made it easier and faster to experiment with cropping and design and provide proofs for the client. The marbled paper was later scanned by a professional color separator, and the color separations were dropped into the film used to make plates for printing the book.

ELECTRONIC MARBLING

What if you just can't lay your hands on a suitable piece of marbled paper? Then start with a scan of any image that distributes color and shape somewhat evenly and apply filters in Photoshop or Painter to create an electronically marbled texture. Starting with a scan of newspaper text, for example, makes it possible to add colors that you define, for projects requiring a particular palette. Starting with a scanned photo or piece of cloth, on the other hand, provides colors that are especially subtle and interesting. The marbling produced by electronic means looks remarkably similar to hand-marbled designs and can be used for the same sorts of projects— for example, gift wrap, decorative boxes, lamp shades, journal covers, package designs, or backgrounds in book design.

STARTING WITH TEXT

We started with a scan of characters from our daily newspaper and filled the type and the background with solid color. We then selected the type again, offset the selection, and filled the selection with another color to create a staggered pattern containing three colors plus the white left behind in the counters of the original type. We applied the Motion Blur filter to smear the edges of the color and then applied the Ripple filter to marble it. Then, to soften the grid-like look of the pattern, we rotated it 45 degrees and rippled it again to produce a final pattern.

USING PAINTER'S MARBLE FUNCTION

Painter, a program dedicated to reproducing the look of natural art media, includes a function for applying a marbled distortion to images. We used it to marble a scan of a piece of paisley cloth and a fairly monochromatic photograph of a succulent plant from a garden.

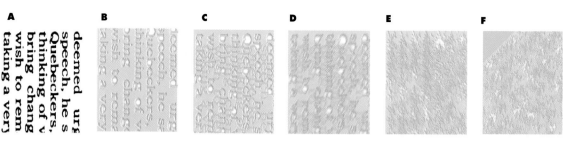

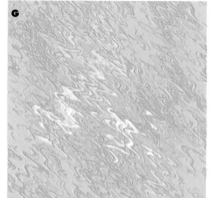

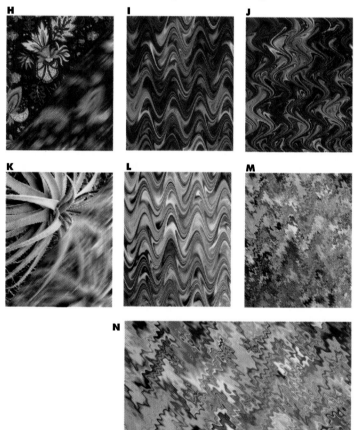

Using Photoshop's Ripple command
To electronically marble a pattern with specific colors, we started with a scan of black-and-white newspaper type, removed the gray tones from the background (**A**), added color to the type and the background, and left the counters (the "holes" in the round letters) white (**B**). We then selected the type, dragged the selection outlines down and to the right and filled this area with a third color (**C**). We applied the Motion Blur filter with an angle of 45 degrees and a distance of 20 pixels to blend the colors and blur the type (**D**). We then applied the Ripple filter with a Large frequency and an amount of 500. The result had a pleasant texture but the vertical orientation of the original lines type was distractingly visible (**E**). To fix this, we rotated the image 45 degrees clockwise (**F**) and applied the Ripple filter again. We then selected a section of the pattern that we liked and enlarged it to fit the size and resolution we wanted. We applied the Median filter to soften the stair-stepping that occurred as a result of the resampling and applied the Unsharp Mask filter to sharpen the edges and improve the contrast of the color areas (**G**).

Using Painter's Apply Marbling command
We started with a scan of paisley cloth, applied a motion blur in Photoshop (**H**) and then used Painter's Apply Marbling command at its default values for spacing, offset, waviness, wavelength and phase, which duplicate the attributes of a marbling rake (**I**). We applied the command again with a contrary motion to produce a final design (**J**). We used the same method to marble a photo of cactus (**K**), but after the first pass with Apply Marbling (**L**) we returned to Photoshop and applied the Ripple filter to vary the design (**M**). In both cases, we applied the Median and Unsharp Mask filters to the final image to eliminate pixelation and improve contrast. A close-up shows the subtlety of the electronically marbled cactus image (**N**).

8 | Working with Scanned Photographs

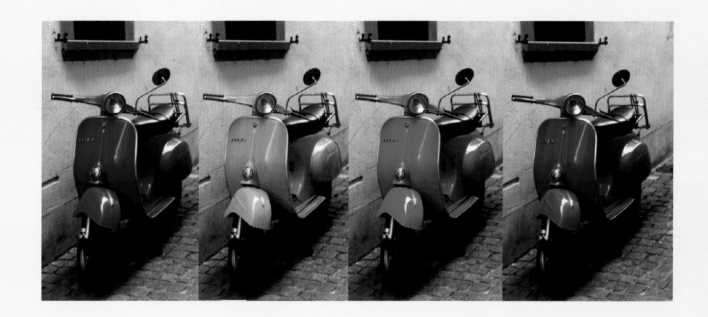

Altering Scanned Photographs

"FIXING" A PHOTOGRAPH

Altering is a nice word for describing a kind of fiddling around with photographs that might also be called "distorting" or "falsifying." Photo altering has been done almost since photography was invented, and some more recent incidents have become worldwide events (for example, the famous altering by *National Geographic* of a magazine cover photo of the Egyptian pyramids). Though such distortions have always been possible, it's much easier to alter photos on a computer than in a darkroom, so the practice is now available to anyone with a desktop system and an image editor.

THE ETHICS OF IMAGE ALTERING

Image editing is a wonderful tool because it frees us from using photographic images straight from the camera. We can "improve" photographs (make them more aesthetically pleasing or better suited to our purposes) by changing certain elements in ways that don't dramatically distort the meaning or veracity of the image—for example, by making the sky more blue, or by cloning over a piece of trash with a sample of the surrounding grass. But we can also change photo images in more dramatic ways that do change their content and meaning and call into question the use of photographs as records of reality. Adding the head of a board member who was absent the day the group shot was taken is a relatively benign use of image altering. Going a step further, some people ask computer artists to delete an ex-spouse from favorite family snapshots. Further still, criminals add or delete people and things from photographs presented as evidence in court. Motivations for image altering range from the convenient to the fraudulent. We'll use some rather harmless examples to show you how to alter what the camera saw, and leave it to you to use these techniques wisely.

ADDING ELEMENTS

One of the easiest ways to alter a photograph is by adding elements taken from another image, especially when the new element is added right on top of the original image. With a little more effort the new item can be layered into the existing image so that it appears in front of some

Adding a figure
Janet Ashford and her friend Hale photographed each other outside a mountain cabin on the same day and at the same hour, and it happened that both subjects were facing the same compass direction and hence were lit by the sun in a similar way. This made the two photos (**A**) good candidates for combining. We scanned the photos and in Photoshop used the lasso tool to select each figure. The selections were saved in channels, such as the one around Hale (**B**). Working with a file copy, we selected Hale and copied and pasted him into the Janet photo (**C**). Then, because Hale is taller than Janet, we used the Scale function while he was still selected to make him larger (**D**). We then positioned the enlarged Hale next to Janet, where he looked fairly comfortable (**E**). But the shadows on Hale's face were a little darker than those on Janet's, so we selected his face (**F**) and used Levels to make the tonal range fit better (**G**). The Dust & Scratches filter and then the Unsharp Mask filter was applied to the final image, which looks reasonably realistic (**H**).

A

Anything more than the truth would be too much.
—Robert Frost

Layering a figure behind another figure

The composite shown on page 76 looks a little awkward, since the man usually stands behind the woman when the two are photographed. So to make our composite more convincing we used our previously saved selection outline to select the original figure of Janet and copied and pasted it into the composite so that it covers the left side of Hale and makes it look as though he is standing behind her (**A**). But now it was necessary to darken the part of Hale that would have been shadowed by Janet if he had really been standing there. We selected an area of Hale (**B**) and used Levels to darken it (**C**). Then, to produce a final version, we painted out the light edge of Hale's pants, darkened his left hand slightly, smoothed the transition between his hat and the sky, applied the Dust & Scratches filter to clean up some scratches, and sharpened the image using the Unsharp Mask filter. The result (**D**) looks remarkably realistic. If this image were output to a glossy print, such as an Iris proof, probably most people would not question its veracity.

B

parts of the original image and in back of others. In either case, it's important that the lighting, color palette, and highlights and shadows of the original image and the new material are similar. Your composite image will immediately look faked if one object has shadows cast by a light source on the left, for example, and another has shadows cast from the right. However, if the images are similar enough, making careful adjustments to tonal range, edge blending, saturation and so on will work to make a combined image look like an unaltered original.

DELETING ELEMENTS

Sometimes a good photo is marred because an extraneous person happened to walk into the background at the wrong time. The most seamless way of correcting this is to use a tool like Photoshop's rubber stamp tool in Clone mode to pick up samples of the surrounding background and use them to paint over the unwanted

C

D

Removing unwanted elements

People walking on the grass during an Easter egg hunt cluttered the background behind a little girl. Cloning the grass with Photoshop's rubber stamp tool made it possible to paint out the unwanted figures. Likewise, a portrait of a sunbathing couple was marred by an extraneous figure and a blue beach bag. Since both the rock and ocean are naturally textured areas with little specific detail, it was easy to sample these areas with the rubber stamp tool and use the samples to paint out the unwanted elements.

Appearances are deceptive.
—Aesop, 550 B.C.

element, so that attention is focussed on the subject. This works especially well if the background is somewhat amorphous (such as ocean water or leafy foliage). But there must be enough background area available for a good sample and the background must be sufficiently simple. Deleting an unwanted figure by painting over it with samples of neighboring grass or water is quite easy, since these textures can be cloned and spread around in a way that looks natural. By contrast, a street scene background contains too much detail and color and too many sharp angles to be cloned easily. Falling somewhere in between is a photo of a musician sitting on wooden steps. In order to delete a figure to his right, we had to carefully build selection areas that outlined the area of the obscured steps without overlapping the neck of the instrument. These areas were filled with wood texture sampled from the surrounding steps.

Redefining the truth

We liked a photograph of a man playing cittern, but wanted to eliminate the distracting figure of a woman sitting near him (**A**). We used the lasso tool to select the figure and then deleted it, taking care to leave the tuning pegs of the instrument intact (**B**). We then used Photoshop's rubber stamp tool to sample areas of the top steps and painted into the blank areas above the instrument to restore the steps (**C**). We clicked in the remaining large white space with the magic wand to select it and saved the selection in a channel (**D**). We then used the lasso tool with the Option key held down to draw straight lines that define the shape of the entire middle step and saved this selection (**E**). We combined these two selections in a third channel to produce a selection that defined only the part of the second step surrounding the instrument (**F**). We then were able to paint into the selection with samples of the step, without painting over the cittern neck (**G**). We used the same procedure to create a selection area for the top of the bottom step and filled it with wood texture sampled from other areas of the steps (**H**). We filled in the remaining white gaps with sampled wood textures to complete the reconstructed steps (**I**).

F

B

G

C

H

D

I

A

E

MORPHING BETWEEN TWO PHOTOGRAPHS

One of the more exotic techniques available for altering photographs on desktop computers is known as *morphing*. Morphing is the blending of two different images in a sequence in which one image appears to turn into another as if by magic. The technical term for this process is a *spatially warped cross-fade*, but in spite of the intimidating terminology it's quite easy to do if you have the right software. For the examples on this page we used the application Morph by Gryphon Software. Morphing works best when the two photographs are similar in orientation and lighting. For example, a morph between two people wearing hats against a dark background will be more successful than one between two people, one with a hat outdoors and one without a hat indoors.

HULTON DEUTCH

Starting and ending images with key points

Freud/Wayne morph
To make a morph between Sigmund Freud and John Wayne we placed corresponding points on the starting and ending images. When a point is selected it turns red in both windows. Points can be maneuvered in each window independently so that specific key points, such as eyes and chins, line up in each image. (Images from Aldus/Hulton-Deutch *Vol 1: People & Personalities* Photo-CD.)

Erosion
Here two rock formations have been paired in a sequence suggesting erosion. Aside from its obvious entertainment value, morphing is a technique that can be used in educational and scientific productions. (Images from Gazelle Technologies *Creative Backgrounds and Textures* Photo-CD.)

Enhancing Scanned Photographs

BOOSTING COLOR

Sometimes a photograph is strong in content and composition but has washed out color, either because the film was overexposed, or the camera was pointing toward the sun, or simply because it was an overcast day. Much can be done in an image-editing program to boost the saturation of such a photo, either in whole or in part. When overall thinness of color is the problem, try increasing the saturation of the entire image. When only part of the image needs correcting—for example, the white sky of a cloudy day—select that part and play with saturation and other controls to deepen the color or change its color components.

CHANGING COLOR

Changing a white sky to blue can change the feeling of a photo from melancholy to cheerful, without significantly changing the pictorial content of the image. It's an edit that won't be noticed by most viewers. But sometimes a more dramatic change in color is desired. Using controls for hue, saturation and selective color, it's possible to change the colors in all or part of an image so that they are completely different from those of the original. This is especially useful in advertising and package design for creating different color versions of the same object—for example, shoes, bicycles, or other items that come in several colors—without having to take several photographs.

Improving color

A photo of an arboretum had good detail and composition but dull color (**A**). We used Levels in Photoshop to adjust the tonal range and then boosted color by increasing the saturation of the whole image (**B**). We selected the sky with the magic wand, increased the saturation of this area, and used Selective Color to increase the cyan component of the sky (**C**). We then increased the saturation of the whole image again and applied Unsharp Mask to sharpen and improve contrast. Type and border elements were added to create a poster (**D**).

A

B

C

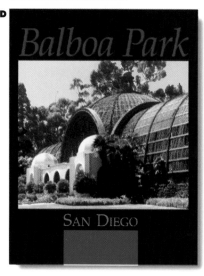

D

A

B

C

D

Changing color

A sleek new Italian motor bike rests on an old cobbled street in the heart of Zurich, providing a characteristic image for a postcard of the city (**A**). To change the green parts of the bike, we used the lasso to select the painted metal and also most of the silver chrome, which contains reflected tones of the surrounding green. The selection was saved in a channel, shown here in Photoshop as an overlay over the RGB image (**B**). Then we used Hue/Saturation to change the color of the bike from green to blue. To create variations in red and yellow, we used Hue/Saturation to slide the color into the desired hue range and then used Selective Color to fine-tune the color components (**C**). The four color variations were cropped and combined in one image and then a panel at the top was lightened and type was added (**D**).

ADDING COLOR TO GRAYSCALE PHOTOGRAPHS

In the days before color photography, portraits and family groups were shot in black-and-white and sometimes printed as sepia-toned prints. These old photos, as well as family snapshots taken through the 1950s, were often colorized by painting directly on the photographic prints with special paints or inks. Hand-tinting added a special, subtle look to old photos, as soft smudges of pink blossomed on cheeks and robin's egg blue appeared—sometimes improbably—in eyes. Some illustrators still use the old-style photo paints to add ethereal color to black-and-white prints. This look can be duplicated electronically to add soft color in the traditional way, or it can be exaggerated to create a special effect that makes an old subject look contemporary.

Adding subtle color

We started with a scan of a photo of Janet Ashford's mother, Alice Munro, taken at UCLA around 1946. We cropped the image, adjusted the tonal range in Photoshop, sharpened the image with Unsharp Mask, and converted the grayscale scan to RGB (**A**). Then, to prepare for tinting, we used the lasso tool to select areas of the image that we wanted to change (skin, dress, hair and shoes, foreground seat and background wall) and saved each selection in a separate channel (**B**). (The selections were used later to create a graphic treatment of the same image, which is described under "Creating Poster-Style Graphics" on page 96.) We then filled each selection area in turn with a different color (**C**), using 20% opacity so that we could gradually build up tone by filling more than once, and using the paint in Multiply rather than Normal mode so that the color was added without greatly changing the black tones of the original. When all the selection areas were filled, we used the paint brush to add color detail to the eyelids, lips, and cheeks (**D**). The result is a very subtly colored portrait of a stylish young woman caught in an unusual pose (**E**).

A

C

D

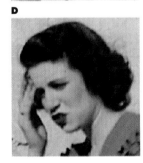

B

E

F

Boosting the color

To create a dramatic color treatment, we started with the softly tinted version and used Variations in Photoshop to view and choose increases in color saturation and CMYK components. This process added color to many of the black areas, so we then used Selective color to restore the blacks by increasing the black component in the black and neutral areas of the image (**F**).

A

B

C

D

CREATING DUOTONES

A duotone—a print using two inks—adds depth to a photograph. Photoshop makes it easy to create duotones, tritones and quadtones. The curves for each color can be manipulated to create strong or subtle effects. However, duotone files may not output correctly when placed in pages created in other programs, so check to see how spot color duotones are handled by the page layout program or color separation utility you're using. Since this book is printed in the four process ink colors (CMYK), we can only simulate the effect of spot color duotones on this page.

E

One of Photoshop's predefined pairs of duotone curves: PMS 527.Bl1

F

Another of Photoshop's predefined pairs of duotone curves: PMS 527.Bl4

G

H

I

J

Spot-color quadtone

One of Photoshop's preset quadtone curves (Black, 541, 513, 5773) (**I**), adds one more ink color but looks similar to a tritone. Curve manipulations (**J**) were designed to shift the peak of each color to a different part of the tonal range, allowing each one to become dominant in some part of the image.

Spot-color tritone

One of Photoshop's sets of preset tritone curves (Black, 165, 457) (**G**), gives a subtle, rich effect to the photograph which is not greatly different from a duotone. Experimenting with the curves (**H**) can produce special effects. Whenever a curve is bent into a wave, the effect on the image is like a solarization, in which the positive and negative values are inverted in parts of the grayscale.

Nearly a duotone

The simplest way to add richness to a black-and-white photograph (**A**) is to print it in all four process colors (**B**). Another simple way to get a duotone effect is to take the black and white photograph and lay a flat tint under it (**C**), in this case, 25% yellow. The problem with this method is that the highlights are replaced with the tint color, flattening out the tonality of the image. A true duotone (**D**) reproduces the tonal range of the photograph in each ink color—in this example, black and cyan.

Spot-color duotones

Duotones can be strong, for example with black reduced in the midtones and purple reduced only in the shadows (**E**); or subtle, with purple suppressed and a normal black curve (**F**).

A

ADDING TEXTURE TO A PHOTOGRAPH

By using the many native Photoshop filters and third-party plug-ins, you can add an almost limitless variety of texture effects to photographs (see pages 35, 66–67 and 90–92). But you can extend the range of possibilities even further by adding your own textures to photographs, or by using textures that have been scanned from clip art sources (see page 56). There are many ways to add texture to an image. Experiment to find the combinations that are most pleasing or effective.

B

Applying a line art texture
The original color scan (**A**) was combined with a scanned line art texture (**B**) to produce a reticulated effect (**C**). To achieve this we pasted the texture into a channel, loaded it as a selection and applied a +175 value to the black setting of the Input Levels (in Photoshop's Levels dialog box).

C

Adding texture to the black channel
To add the texture of a drawing to the photograph, we converted the RGB image to CMYK and pasted a scan of pencil shading (**F**) into the black channel only (**G**). The pencil texture was pasted into the black channel with "darken" as the composite control option (**H**).

D

Laying a texture on top
Using a texture scanned from screen door mesh (**D**) as a channel, the photograph is broken up with a pattern that goes from black to white horizontally across the image in a graduated fill (**E**). In this case the texture lies on the surface, rather than blending in.

E

F

I

G

H

J

Pasting into a texture
The irregular outline of a piece of handmade paper (**I**) frames a crop of the original photograph. Here the photo has been pasted onto the texture, rather than the other way round, as in the other examples. To select the paper shape, we first selected the background outside the edge of the paper, then inversed the selection. The composite or floating selection mode was Multiply, causing the superimposed image to darken the underlying gray paper (**J**).

A

B

C

MEZZOTINT EFFECTS WITHOUT USING FILTERS

A mezzotint is a screening method applied to tonal images that breaks them down into random patterns of irregular dots. The texture is reminiscent of the old lithography technique of drawing on stone with a special crayon. In pre-digital days we would often use a mezzotint to "save" a hopelessly fuzzy or otherwise flawed photograph. In the digital world an analogous process is known as "dithering." It isn't exactly the same as a mezzotint but it's awfully close. Dithered images were the only kind that the very first scanners for desktop computers could produce, so when more sophisticated grayscale and color scanners were introduced, designers tended to ignore dithered images. Although Photoshop's filters provide various mezzotint effects, we usually get better results by using dithering.

D

E

 Enlargement of a digitally dithered image.

 Enlargement of a traditional mezzotint.

Mezzotinted or dithered bitmap files are handy in many ways: They occupy little disk space and print quickly, they can be combined in transparent layers in a layout program (see page 44), and they can *still* save that awful photograph.

Mezzotint effects can be applied in color as well as black-and-white. A few of the many possible ways are shown on this page.

F

G

Creating a monochrome mezzotint

The original color scan (**A**) was converted to grayscale mode (**B**) and then converted to bitmap mode with a diffusion dither (**C**). This produces an effect not unlike a standard graphic arts mezzotint.

Producing a color mezzotint

After establishing a bitmap mode file the same size as the image, each of the cyan, magenta and yellow channels in (**D**) was copied and pasted into the bitmap mode file and then copied back into the color channel as a dithered bitmap. The black channel has been left in continuous-tone. The same process was applied to an RGB file (**E**), with all three channels dithered.

Trying variations

For one variation, the three color channels of a CMYK file were converted to bitmap mode with 50% threshold instead of dithering and only the black was dithered (**F**). For another version, all three channels of an RGB file were converted to bitmap mode with the 50% threshold option (**G**).

9 | Transforming Photos into Graphics

Bitmapped Effects with Photographs
Creating High-Contrast Images
Posterization
Solarization
Applying Graphic filters to Photos
Creating Montage Illustrations

PostScript Effects with Photographs
Creating Poster-Style Graphics
Creating a Silk Screened Look
Applying Custom Line Screens

Bitmapped Effects with Photographs

CREATING HIGH-CONTRAST IMAGES

High-contrast images are dramatic and fascinating. They reduce the continuous tones of a photographic image to areas of stark black and white that emphasize only the most crucial elements of a picture. The effect is especially striking with human faces. The shapes that make up eyes, nose and mouth—shapes that our brains are hard-wired to search for and recognize—suddenly stand in sharp contrast to the field of white around them, exaggerating expression and evoking a strong emotional response.

High-contrast images have been created in the conventional darkroom for many years and have become part of the visual vocabulary of design and illustration. Creating them with desktop computers is relatively simple. The key is to adjust the midtones, as well as the light and dark ends of the tonal range, so that as contrast is increased there remains enough visual information to make the image readable.

Continuous-tone images can be converted to high-contrast versions using the tonal range functions available in most image-editing programs. One can also use certain filters to convert a photographic image to a drawing-like image that contains only black and white. In addition, it's possible to create special effects by layering a high-contrast version of an image over a copy of the original photo.

A

B

C

D

E

Producing a high-contrast image in stages
This woman's dramatic expression makes a good subject for conversion to a high-contrast image. But the color photo portrait was taken in shade and does not have the deep shadows that exaggerate features and make high-contrast conversion easier (**A**). We converted the image to grayscale mode (**B**) and carefully adjusted the midtones using Levels in Photoshop, editing the photo in stages (**C**, **D**) until we were able to use the Brightness/Contrast controls to produce a version that is almost entirely black and white (**E**).

 A

TRANSFORMING
PHOTOS INTO
GRAPHICS

87

A

B

Using a high-contrast filter
Many of the filters used with Photoshop convert color or grayscale images into graphic black and white. We started with a photo of a trench-coated man at Coit tower in San Francisco. We converted the image to grayscale (**A**), lightened the midtones (**B**), and applied the Charcoal filter from Gallery Effects 1, which changed the gray tones into black strokes on white (**C**).

C

USING A HIGH-CONTRAST IMAGE AS AN OVERLAY

To get the drama of a high-contrast image, but still retain the rich texture of a continuous-tone photograph, we scanned a photo of an Italian castle, saved a copy of the original (**A**), converted the color image to grayscale, and then used Levels and Brightness/Contrast in Photoshop to produce a high-contrast version with no gray values at all (**B**). We selected the black areas of the high-contrast version, copied the selection and pasted it back into our copy of the original (**C**), being careful to line up the selection so that it fit the underlying image. We then saved the selection, and inversed it to select everything *except* the black areas. We used Hue/Saturation to boost the saturation of the selected areas so that the sky looks dramatically stormy and the castle walls have the warm look of sunset. These textured areas are set off by the high-contrast black, which defines the architectural structure, adds impact and also helps obscure the figures of tourists on the castle bridge. The technique transforms a rather ordinary photograph of an historical building into a memorable image.

D

When the Many are reduced to One,
to what is the One reduced?
  —Zen koan

POSTERIZATION

The posterization process reduces the number of tonal levels or levels of brightness in a continuous-tone image, so that instead of smooth transitions from tone to tone, there are sudden jumps between tone levels, producing areas of flat color. For example, when a photographic image is scanned in grayscale mode with 256 shades of gray, and then converted to a 4-level posterization, the posterized image contains only 4 different shades of gray. Posterization dramatically converts a photo into a "graphic" by converting areas of photographic detail into tone-filled shapes that are more similar to what a graphic artist might produce when rendering a scene with ink wash or crayon. Posterization is done in the darkroom by using color filters or by varying exposure times to create a number of different negatives, each of which contains areas of a single tone. These are combined into a master negative and printed. Photoshop makes the process easy by providing a Posterize command. With it, the brightness value of every pixel in the original is analyzed and converted to the nearest of the brightness levels you specify. Any number of brightness levels between 2 and 255 can be specified and will be equally spaced between white and black. The fewer the number of levels, the more dramatic the results.

When a color image is posterized through Photoshop, the results are a little different than for grayscale images. The brightness value of each pixel is changed, but its hue is not, so there may be only four levels of brightness, but many more than four "colors" (see "Posterizing an Object Scan" on page 114). Nevertheless, posterizing a color photo can create very interesting and graphically satisfying effects.

A

B

C

D

**Posterizing a
continuous-tone image**
We scanned a photo of fruit and converted it to grayscale mode. The gradient bar below the photo shows the 256 levels of gray contained in the scan, which appear to blend smoothly from black to white (**A**). We used Photoshop's Posterize command to specify a 4-level posterization, which converted both the image and the gradient bar to four gray levels (**B**).

To show how posterization affects a color image, we added a rainbowlike bar (copied from one of Photoshop's color palettes) with smooth transitions between the colors (**C**). We then specified a 4-level posterization. The myriad colors in the original have been converted to a smaller number of hues that share four brightness levels (**D**).

A

Coloring and simplifying a posterized graphic
To create a graphic for a library poster, we began by scanning a photo of a girl reading (**A**). We edited the color scan to improve tonal range and contrast and silhouetted the figure by removing the bookcase background (**B**). We then used Photoshop's Posterize command to convert the image to a 6-level posterization. The resulting image has a graphic look, but the colors are somewhat inharmonious (**C**). To improve the graphic qualities and color of the image, we converted the edited color scan to grayscale (**D**), and reduced contrast so that there were fewer gray tones in the image. We then created a 6-level posterization (**E**), converted the image back to RGB mode, selected the shapes that had been created by the posterization process and filled them with colors drawn from a more pleasing palette. In some cases we eliminated shapes to make the graphic simpler. The illustration was combined with type to produce a poster (**F**).

B

C

D

E

F

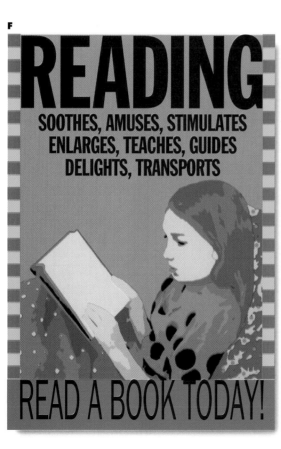

SOLARIZATION

When a photographic print or negative is briefly exposed to light during development, the result (called a solarization) is a blend between the negative and positive image. Solarization creates an eerie effect that can be used when you want an image to look photographic but "unreal." Other interesting effects result from blending positive and negative images using different modes in Photoshop.

A

B

C

D

E

Blending positive and negative
We scanned a photo (**A**) and solarized it using Photoshop's Solarize filter (**B**). This filter leaves half of the brightness values in the image normal and inverts the other half, analogous to the darkroom technique. To create a custom solarization we used the Curves controls to edit the image curve, producing a dramatic effect in the image (**C**).

Another interesting way to produce a solarization effect is to combine negative and positive versions of an image. We inverted a copy of the original to produce a negative (**D**) and then combined it with the original positive image using Photoshop's Difference mode, with the negative image blended at 60 percent opacity. This reversed the hair tones and made the portrait look out-of-the-ordinary, but because the flesh tones are still in the yellow-red range, the face is readable (**E**).

APPLYING GRAPHIC FILTERS TO PHOTOS

filters are sets of calculations designed to apply special effects to images. Photoshop includes a number of native filters, which are grouped into categories that include Blur, Distort, Noise, Pixelate, Render, Sharpen and Stylize. In addition, a number of software developers have created filters than can be used with Photoshop. The filters found in the three Gallery Effects collections are among the best for transforming scanned photographs into images that have interesting graphic qualities, but without overly distorting the content of the photo. Most of the Gallery Effects filters shown here were used at their default values, but these filters, as well as some native Photoshop filters, include two or three parameters (for example amount, edge intensity, cell size, light and dark balance) that can be adjusted to produce different effects. (In Photoshop, one can also alter the effect of a filter by applying it when a channel containing a solid gray tone is selected, so that the filter works at a percentage of its full strength.) By adjusting parameters, or by using a mask, or by applying several filters in succession, very rich variations can be created.

Original image

Photoshop, Add Noise

Photoshop, Crystallize

Photoshop, Emboss

Photoshop, Find Edges

Photoshop, Unsharp Mask

Photoshop, Trace Contours

GE 1, Chalk and Charcoal

GE 1, Charcoal

GE 1, Dark Strokes

GE 1, Dry Brush

GE 1, Emboss

GE 1, Fresco

GE 1, Poster Edges

GE 1, Smudge Stick

GE 1, Watercolor

GE 2, Accented Edges

GE 2, Colored Pencil

GE 2, Grain

GE 2, Glowing Edges

GE 2, Notepaper

GE 2, Palette Knife

GE 2, Photocopy

GE 2, Rough Pastels

GE 2, Stamp

GE 3, Conte Crayon

GE 3, Crosshatch

GE 3, Cutout

GE 3, Ink Outlines

GE 3, Paint Daubs

GE 3, Sponge

GE 3, Sumi-e

GE 3, Watercolor

Using both versions
To create a poster for a series of art classes we combined the original photo of a girl with the version rendered by the Gallery Effects 2 filter, Colored Pencil.

SPECIAL EFFECTS WITH FILTERS

There are many ways to expand on the repertoire of ready-made filters to achieve very special effects. You can try applying a filter to only one channel of an image, for example, or combine a filtered image with a copy of the unfiltered original using one of Photoshop's blending modes, or apply more than one filter to the same image. It's also possible to apply a filter when a channel containing a gradient is active, so that the strength of the filter is affected in a gradual way across the image.

To create an interesting color variation we applied the Gallery Effects 2 Colored Pencil filter to the Blue channel of the image.

For another look, we converted from RGB to CMYK, increased saturation and lightness, and applied the Colored Pencil filter to the K plate.

This graduated mask from black to white was created by using Photoshop's Gradient tool in linear mode.

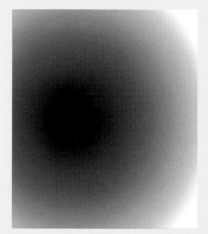

This graduated mask was created by using Photoshop's Gradient tool in radial mode.

We converted the original image from RGB to CMYK and applied the Dust & Scratches filter at 6 pixels to soften the composite image. We then pasted a previously saved version of the original K channel into the K channel of the edited version and applied the Photocopy filter to the K channel alone.

To create this inverted image, we increased saturation in the original and applied Photoshop's find Edges filter. We then opened a copy of the original and converted it to CMYK. We applied the find Edges filter to the K channel only, copied this channel, and pasted it into the working image in Darken mode.

To produce this image, we created a linear gradient from white to black (shown above) in Channel 4 of the RGB Photoshop image, made it active by choosing Load Selection from the Selection menu, and applied the Colored Pencil filter.

To make a variation, we created a radial gradient from black to white (shown above) in Channel 4 of the RGB image, made it active, and applied the Glowing Edges filter.

CREATING MONTAGE ILLUSTRATIONS

A *montage* is an illustration in which realistic components are combined to form an abstract design. Objects may be out of scale with each other or may form unusual spatial relationships. The juxtaposition of images in unexpected, bizarre ways can produce the quality of a dream, in which seemingly incongruous images coexist. It's left up to the beholder to unravel the meaning.

The typical film poster is often a montage, in which different scenes and characters are combined, often at different scales. Since we are accustomed to the idiom of montage we tend not to be confused by a miniature mounted posse, for example, riding along the brim of a gunslinger's hat.

A montage of loafers, pubs and street musicians
Part of the background and all the red areas in the picture have the engraved map pasted into them. The lines surrounding the fiddler are caused by stroking the selected silhouette with white.

Making a digital sketch
You can work out your composition by layering thumbnail-sized bitmap images. Use this layout as a template for assembling the final collage.

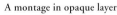

A montage in opaque layers
The unifying theme in this set of photographs taken in London's Portobello Road is color: reds, blacks and oranges. You can apply the same principles of aligning shapes for opaque layering as for transparent layering (see pages 94–95), but the order of layers is more important since anything underneath will be covered up completely. It's a good idea to silhouette at least one of the images on the upper layer.

USING TRANSPARENT LAYERS

There are two basic ways to approach a montage. The images can be opaque and overlapping (*decoupage* is another name for this style), or they can be in transparent layers, much like a double exposure onto film. The first step in creating a montage in transparent layers is to find a set of photographs that seem to go together. It may be a common theme, such as buildings, animals, people or boats; it might be color, texture or just a mood. To keep it simple, limit the originals to three or four.

Having made your choice, now completely ignore the meaning of the images and look for rhythm and shape. Make thumbnail-size line versions of your scans, turn them upside down, rotate, crop and scale them until the composition begins to take shape. Look for matching alignments, such as horizons or windows, and play with butting and overlapping elements so that one set of shapes flows into another. Let the outside boundaries of the cluster of images be irregular if necessary, rather than force everything into a tidy, square frame. Allow accidental phenomena, like crooked scans or torn edges to influence the direction the montage takes as it develops.

Using the thumbnail as a guide, proceed with the full-size montage. In the example shown here, all the images have been superimposed with Darken mode selected.

Making a thumbnail
The only thing that the photographs above have in common is that they were taken in the same year at the same location.

In the thumbnail layout of the montage, elements have been overlapped, rotated and scaled to create a balanced composition. Schematic color has been applied to show the order of layering.

Translucent layering

Following the rough layout of the thumbnail on the opposite page, we built the montage layer by layer. Each source image was slightly rescaled before pasting into position. The black background was kept on a separate layer until the composition was finished. The fuzzy edge of the background shape was made with a combination of the Diffuse and Gaussian Blur filters.

In the days before computers it would have been very difficult, if not impossible to use darkroom techniques to create the subtle blending of images that is now so easy to produce.

PostScript Effects with Photographs

CREATING POSTER-STYLE GRAPHICS

What's the difference between a photograph and a graphic? Well, even though a photograph is a symbolic, two-dimensional representation that only *refers* to objects, photographs usually look so realistic that they seem the same as the view in front of our eyes. Photos are usually intended to duplicate reality. A graphic, on the other hand, is a more abstract visual image that functions as a symbol or icon, which refers either to real objects or to ideas. A graphic, because it is not strictly realistic, must be rendered in some sort of style—impressionistic, decorative, rough, fine—through which objects are suggested, rather than rendered hair by hair. For millennia, artists have looked at the reality around them and rendered it in visual symbols and graphics. As we've seen, the computer makes it easy to create a graphic symbol by starting with a photographic image and performing calculations on its component color and gray values so that the reduction of nature to symbol is automated in a way. The conversion from photo to graphic is especially striking when photos become the basis for hard-edged drawings. We used Photoshop to posterize an old photograph, then autotraced it and developed it further in a PostScript illustration program to produce an image that is stylized, abstract and *graphic*.

A

B

C

D

From photo to graphic

While adding color to a grayscale photo of a young woman (see page 81), we created selections and saved them in channels. When these selections were overlapped in Photoshop's window, the result was an attractive, silhouette-like version of the photo image (**A**). To create a graphic based on this style, we used the channels to fill each selection with a tone of solid gray that completely painted over the photo image beneath. This new image was autotraced in Streamline to produce PostScript shapes (**B**). In addition, we increased the black tones in a copy of the original image and used the magic wand to select the black areas. These were saved as a selection and also autotraced to produce a very high-contrast version of the photo (**C**). Then, to create a color version, we opened the autotraced posterization in Illustrator and added solid color (**D**). The result was an illustration similar in style to the posters produced by the Beggarstaff brothers around the turn of the century. As a variation, we tried filling some of the shapes with gradients (**E**). finally, we copied our autotracing of the high-contrast black shapes and pasted it over a version with gradient-filled background shapes (**F**).

E

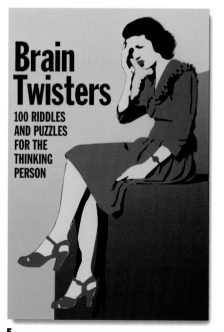

Brain
Twisters

100 RIDDLES
AND PUZZLES
FOR THE
THINKING
PERSON

F

A

B

C

D

CREATING A SILK SCREENED LOOK

To create stylized travel posters of Switzerland, artist Otto Baumberger used paint and a good eye to reduce the broad tonal range he saw in the Swiss landscapes to dramatic areas of flat posterlike color. We took a short-cut to a similar result by using a photograph to create a PostScript posterization that imitates Baumberger's 1930s style. Using Photoshop, we scanned the photo and reduced it to a four-level posterization. We then used Streamline to autotrace the posterized bitmap in Outline mode and opened the autotracing in Illustrator to add color and type.

One of the keys to this technique is careful preparation of the bitmap to separate any shapes that are contiguous. Before autotracing, we carefully checked the bitmapped image to find areas where one shape was touching another—where the gray of the sail touches the gray of the cloud behind it, for example—and drew lines to separate these areas so that the autotracing would produce separate shapes that could be filled with different colors. Though we started with a rather lackluster photo, the resulting PostScript illustration is handsome and looks much like a hand-pulled silk screened poster.

E

F

G

Making a poster

We started by scanning a slightly overexposed photo of a sailboat on Lake Luzern. We converted the color photo to grayscale in Photoshop (**A**). Then we adjusted the tonal range, tried the Posterize command set to four gray levels, readjusted and reposterized until we got a posterization that reduced the photo's information to clearly delineated areas of sky, mountains, boat, and water (**B**). We used the line tool to draw lines separating elements that shared the same gray value (the sail and clouds, for example) so that these would be autotraced as separate shapes (**C**). We then autotraced the image in Streamline as a four-level posterization (**D**). (Though Streamline can posterize a continuous tone image, posterizing first in Photoshop allows more control over the conversion to four gray levels.) To add color, we opened the autotracing in Illustrator, selected each shape and filled it with solid color. We finished our Swiss-style poster by adding red and black type set in Helvetica Condensed (**E**). The finished electronic poster looks quite similar in style to one produced traditionally by Baumberger in 1935 (**F**). As a variation, we also created a single-color version by selecting shapes that shared a gray tone and filling them with similar tints of blue (**G**).

APPLYING CUSTOM LINE SCREENS

Custom line screens can be applied to any grayscale TIFF image. There are two basic ways: using Image Control settings within PageMaker, or using a Photoshop filter module. Images edited with these two methods behave somewhat differently. The Image Control-modified pictures retain their specified screen frequency even if the picture is subsequently rescaled. The screen frequency of the TIFFs modified in Photoshop will scale up or down with the image.

Custom line screens are effective where low resolution printing is needed, such as for cardboard boxes.

A

B

C

D

E

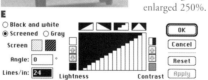

F

G

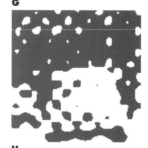

H

Using Image Control
The original scan (**A**) must be converted to grayscale for custom screen work. For (**B**) the Image Control settings were: screen angle 90°, 24 lines/inch; for (**C**) screen angle 0°, 24 lines/inch; for (**D**) screen angle 45°, 48 lines/inch. for (**E**) screen angle 45°, 48 lines/inch, enlarged 250%.

Using Photoshop filters
Accessible from within PageMaker 6.0, Photoshop filters give more options than Image Control, including circular dot (**F**) and line effects (**G**). All the custom screened images can be colored from the palette of the layout program (**H**).

10 | Creating Type Treatments

Using Old Typefaces

START WITH A SCAN

100

TYPE FROM HISTORICAL SOURCES

One important use of the desktop scanner is as an essential adjunct to your existing font resources. Type can be scanned from your original drawings or from old type specimen books. Unlike artwork, typefaces are not protected by copyright, much to the dismay of type designers. Only the *name* of the typeface is protected! You may not scan Helvetica, for example, convert it to a font and market it under the name of Helvetica, although you could if you called it "Swiss Sans" or some such name.

There are two basic ways to handle scanned type: Either use letters as graphics for initial caps or short headlines, or generate a font using a typographic program so that you can type the font characters from the keyboard.

INITIAL CAPITAL LETTERS FROM THE RENAISSANCE

Two examples of the letter "M" by Albrecht Dürer, 1535 (**A**), and an alphabet drawn by Geoffrey Tory in 1529 (**B**), show attempts by Renaissance typographers to reveal the underlying geometry of the letterforms created by the Roman stonecutters. A similar 16th Century letter by Pierre le Bé (**C**) forms the basis of a colored drop cap for a page layout (below). Adding embossing and paper texture (**D**) gives another dimension to the same letter.

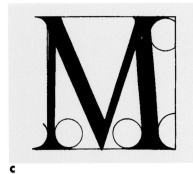

C

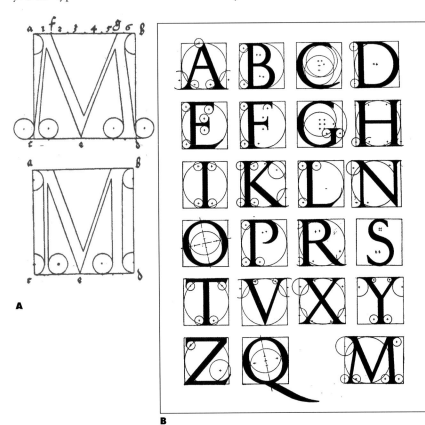

A

B

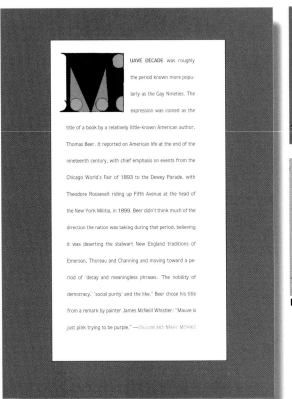

MAUVE DECADE was roughly the period known more popularly as the Gay Nineties. The expression was coined as the title of a book by a relatively little-known American author, Thomas Beer. It reported on American life at the end of the nineteenth century, with chief emphasis on events from the Chicago World's Fair of 1893 to the Dewey Parade, with Theodore Roosevelt riding up Fifth Avenue at the head of the New York Militia, in 1899. Beer didn't think much of the direction the nation was taking during that period, believing it was deserting the stalwart New England traditions of Emerson, Thoreau and Channing and moving toward a period of 'decay and meaningless phrases.' The nobility of democracy,' 'social purity' and the like." Beer chose his title from a remark by painter James McNeill Whistler: "Mauve is just pink trying to be purple." —WILLIAM AND MARY MORRIS

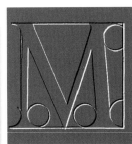

D

CONVERTING SCANNED LETTERS TO POSTSCRIPT FONTS

Of the thousands of typefaces designed since Gutenberg, many are suitable for scanning and converting to computer fonts. But since the process is tedious, it's worth checking to see if the historical type you want is already available from a font catalog.

CONVERSION BY AUTOTRACING

Most font-creation programs have an autotracing feature that instantly converts a scanned letter to an outline (the demonstration fonts in this chapter were made with Fontographer). The process is imprecise, however, and works best with fonts that in their original form have irregular outlines. With autotracing, forming each letter takes a only few seconds; preparing and importing each scanned letter can take hours.

Autotracing a font
Block Condensed (**A**), an alphabet from the early 1900s, is one of many interesting typefaces in *Condensed Alphabets* by Dan X. Solo (Dover, 1986). Because all the corners are rounded and the edges are slightly wavy it's ideally suited for autotracing.

For autotracing, the scan need only be at 72 ppi at about a 1-inch height. Each letter must be copied to the Scrapbook or Clipboard individually and pasted one by one into the template layer of the font program (**B**). If the selection rectangle for each copied character is the same size, then letters will not need to be rescaled in the template.

Kerning pairs must be specified (**C**) to ensure that different letter combinations fit together properly at headline sizes. The standard space around each character, known as the *metrics*, is also specified in this window.

ABCDEFGHIJK
LMNOPQRST
UVWXYZ
abcdefghiijklm
nopqrstuvwxyz
(&;!?"'$)
1234567890

A   **B**

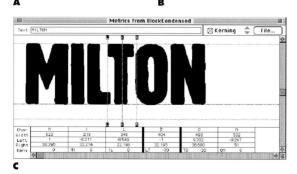

C

Out on the steppes of Russia in January you wouldn't want to be caught dead in a small car.
Literally.
That's why all Russian cars are really big. Especially the Ovlov.

And, of course, all Ovlov models come equipped with the world's best heater.
So don't settle for a car that just looks like a tank. Buy one direct from the tank factory.

O V L O V

CONVERSION BY HAND-TRACING

Hand-tracing—that is, using hand-and-mouse to drive software tools to trace the outline of a letter—is the best method to use when the original letters have sharp, precise edges, straight lines or delicate serifs. It's also your only recourse for the creation of missing letters if your source font is incomplete.

Any letter shape can be broken down into straight line segments and curves that can be constructed with the font-creating software in much the same way as with any other PostScript drawing program. Creating a font this way is an exacting process requiring patience and an ability to focus on details. Stroke widths have to be constant. Each letter has to be checked against every other for consistency. Even a relatively simple san serif typeface with mostly straight lines is a challenge. A serif font involves more attention to detail and requires even more careful checking for consistency.

On this page we have taken a partial showing of a typeface that was created around 1905, extrapolated the missing letters and added punctuation. To make this a complete font, numerals, lowercase characters, accents, special characters such as "$" and "%," and ligatures ("œ" and "fi," for example) would all have to be designed.

Hand-tracing a font

To create clean, blocky type for a poster we started with lettering scanned from an ad for a Berlin advertising typesetter, made around 1905 (**A**). The way in which the "ST" and the "CK" combine is intriguing. Variations in the letters suggest that the original might have been hand-lettered.

Scans of the letters (**B**) were too fuzzy to autotrace. Besides, the nature of this typeface implied crisp, precise corners, making it a perfect candidate for hand-tracing.

Constructing each letter by hand (**C**) involved placing each Bezier control point precisely. Note that the outline deviates from the template to create a more unusual ending to the right foot of the "K," inspired by the word "DRUCK" in the original. More than half of the characters in the font (**D**), including all the punctuation, had to be designed from scratch. The finished font was used to create type for a new poster (**E**).

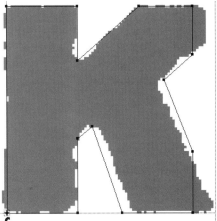

Creating New Typefaces

CREATING AN ORIGINAL CAPS FONT

Creating a clean, well-designed new typeface is an ambitious undertaking and really cannot be achieved successfully in one session. But sometimes a font that's intentionally sloppy is just what you need. Deliberately weird and marginally readable fonts have found acceptance recently, especially in youth-oriented media such as MTV and skateboarding magazines. In this spirit, we created a "grunge" alphabet by using an X-Acto knife to cut rough half-inch high letters from a sheet of black paper. Only straight cuts were used and the strokes were kept to the minimum necessary to form each character. The paper letters were arranged on the scanner with plenty of space between them and scanned at 72 ppi at 400%. After selecting and cutting each letter to the clipboard and pasting the scanned letters into the font software's template layer, we autotraced each letterform to preserve the ragged edges of the original.

A

B

Starting with a pencil sketch
A pencil sketch for a logo often
can suggest an entire font. After some more pencil roughs,
the basis of a typeface begins to
emerge.

SCANNING PENCIL DRAFTS FOR A NEW FONT

Experiments with redrawing historical fonts may give you confidence to try designing a typeface from scratch. A complete typeface is quite complex, especially if it includes both upper and lowercase characters, punctuation, numerals, ligatures, accents and special symbols. Then there are the different weights and italics to consider, to raise a complete type family.

Many font ideas begin as doodles, or as offshoots of logo development. (Herb Lubalin's Avant Garde font began as a magazine nameplate.) A sketch of key characters can be scanned to form a template. The lowercase "h" is a good starting point. It's relatively easy to draw and it determines the relationship of the thick and thin strokes, the ascenders and the x-height. With cutting, pasting and modifying, the "h" forms the basis for the lowercase "n," "i," "j," "l," "m," "n," and "r." The next letters might be "d," "b," and "q," from which the "o," "a," "e" and "c" can be derived. Some letters are tricky: the "g" and "s," for example, are full of exacting curves.

After the whole alphabet has been developed, extensive testing for consistency, spacing and kerning is necessary before your font is ready to leave the nest. Because there are so many fonts on the market, be sure to check your font design software manual on how to avoid conflicts with other fonts. Always include your special font when sending files to your service bureau; better yet, convert all but the smallest type to outline.

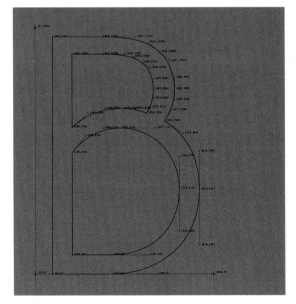

Checking the specifics
A Fontographer printout of a specific character gives detailed information about each Bezier curve and point. To ensure smooth lettering, use as few points as possible to form each character.

From pencil to printouts
As the new typeface evolved, the oblique endings on some of the vertical strokes were abandoned because they impaired legibility.

The still incomplete font was tested at different weights. "Hamburgerfonts" is a traditional test word in type design.

Hamburgerfonts
Hamburgerfonts
Hamburgerfonts
Hamburgerfonts
Hamburgerfonts
Hamburgerfonts
HAMBURGERFONTS

BISTRO

AaBbCcDdEeFfGgHhIiJjKkLl
MmNnOoPpQqRrSsTtUuVv
WwXxYyZz
1234567890
fiflœß

1234567890-=
qwertyuiop[]\
asdfghjkl;'
zxcvbnm,./
!@#$%^&*()_+
QWERTYUIOP{}|
ASDFGHJKL:"
ZXCVBNM<>?

The Bistro font
The complete font includes two weights. (Bistro, designed by John Odam, is available from TreacyFaces, West Haven, CT.)

A FONT FROM SCANS OF YOUR OWN HANDWRITING

If you are looking for a truly original script font, you may already have designed it. Your hand-writing is a typeface that has evolved over a lifetime and is uniquely expressive. To get your writing from pen and paper into font form via the scanner is relatively straightforward using a program such as Fontographer.

Here are some considerations: Put plenty of space between the letters when you write your sample so that each letter can be selected separately (don't attempt to join the letters). Remember to include punctuation. Don't fuss with the letters after they have been autotraced, trying to make them "perfect." Leave the quirks intact.

B

After an alphabetical sample of handwriting had been line-scanned, the letters were selected and pasted one by one to the Scrap-book (**A**). An image from the Scrapbook was pasted into the template layer of a character window and autotraced (**B**). The process was repeated until the font was complete (**C**).

C

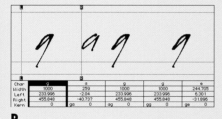

D

In Fontographer the only part of the process of converting handwriting to a font that requires skill, judgment *and* good luck is the determination of the metrics, or the amount of lateral space around each letter. Because not all letters are the same width (an "m" for instance is much wider than an "l"), each one must be treated separately. The initial state of a letter, "g" in his case, has wide spacing around it (**D**). (The spacing of the other letters has already been set.) By dragging the right-hand vertical to the left (**E**), we established the correct distance to a following letter. Dragging the left-hand vertical line to the right (**F**), set the standard distance for a preceding character.

E

A

F

"And this our life, exempt from public haunt,
Finds tongues in trees, books in the running brooks,
Sermons in stones, and good in everything."
—William Shakespeare

A FOUND-OBJECT ALPHABET

Although type design is an exacting task, it can also be a form of entertainment—an artful exercise in visual scavenger hunting. Imagine, almost a complete font is lying in the drawer of your desk. All you have to do is scan it!

Found objects as letters are useful in logos and can make interesting initial caps. A complete alphabet with lowercase, numerals and punctuation would be almost impossible and not really practical since one could hardly imagine wanting to read a page of 10-point found-object type.

For more ideas and tips on scanning found objects, see Chapter 11, "Scanning Real Objects."

The shapes of many household objects correspond roughly to the letters of the alphabet.

A scanned object could form the basis for a tightly drawn alphabet

Paper clips have great potential as numerals.

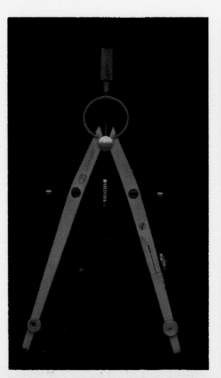

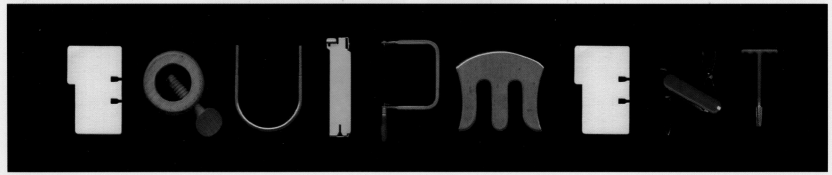

11 | Scanning Real Objects

Bypassing the Camera

"FOUND ART": SCANNING REAL OBJECTS

Sometimes the source of an original illustration is as close as that cluttered catchall drawer in your kitchen. Any object that's small and light enough to fit on the glass of your scanner can become an instant photo image, to be used on its own, or as the starting point of a refined illustration. By changing colors, applying special filter effects, or converting scanned images to line art, posterizations or duotones, you can create a variety of illustrations that are original, cheap, and readily available. (For more examples of scanning objects see "Textures All Around Us" starting on page 62).

The scanner as camera
All the images on these pages were created by placing small items directly on the scanner glass. Each object was silhouetted in Photoshop.

Of course, flatbed scanners were designed for scanning images on flat paper. But that restriction never stopped curious employees from photocopying their hands and faces on the first office copiers, and inventive designers were quick to see the potential for scanning objects when the first desktop scanners hit the market. Experimentation and an inventive mind can produce unique images that break away from the "period" look of 19th Century clip art or the prepackaged look of more contemporary clip art offerings.

TIPS FOR SCANNING OBJECTS

Scanning everyday objects is fun, but it requires some special attention.

FOCUS

With objects that have protruding parts—for example, a ceramic mask with a large nose—the background parts will be darker and less focussed than the parts that are closest to the glass. You may be able to compensate for this with image-editing tools, or you may want to stick with objects that are flatter.

BACKGROUND

Setting a very thick object on the scanner may prevent you from closing the scanner lid. The scanned object will then have a dark background, with the degree of darkness determined by the amount of ambient light in the room. You can edit out a dark background in an image-editing program (see "Silhouetting Scanned Objects" on page 110). Or you can construct a temporary reflective "lid" for the object by holding a piece of white-foam core board over it.

SHADOWS

Objects will cast shadows as the scanner's light source passes across them, but these shadows don't look the same as those cast by a directional light. To achieve a more natural looking shadow, silhouette the scanned object and create a soft drop shadow in your image-editing program (see "Creating Those Nice Soft Drop Shadows" on page 111).

UNWIELDY OBJECTS

Objects that are round or unbalanced may topple over or roll around on the scanner glass. Keep handy a supply of artboard, an X-Acto knife and tape to construct armatures and restraints for your more unwieldy subjects, to keep them still while they're scanned.

METALLIC OBJECTS

Because the three colored light beams (red, green and blue) from color scanners emanate at different angles, they may cause metallic objects to appear exotically multicolored when scanned. You may like the look of a rainbow-hued monkey wrench. But if not, converting the scan from color to grayscale solves the problem. Silver or steel surfaces look quite realistic in black and white. You can add a monochrome color back to a grayscale scan to achieve the look of gold or copper. If parts of the object are nonmetallic or colored, select only the metallic parts and then reduce the color saturation of these areas to zero.

KEEPING YOUR SCANNER CLEAN

Scanners contain electrical components and are not designed for scanning liquids, though you could probably scan a fresh tomato slice if you're careful and ready to mop up the juice. Likewise, it would not be a good idea to scan dirt or flour or other fine particles that could clog the scanner's works. But we have had success scanning small things like seeds and popcorn that generate some chaff. Carefully remove such items after scanning, use a portable vacuum to suck up the grit, and clean the glass with window cleaner to keep your scans dust-free. It's also possible to lay wet items—including painted brush strokes or ink drawings—on a piece of clear acetate placed on top of the scanner glass.

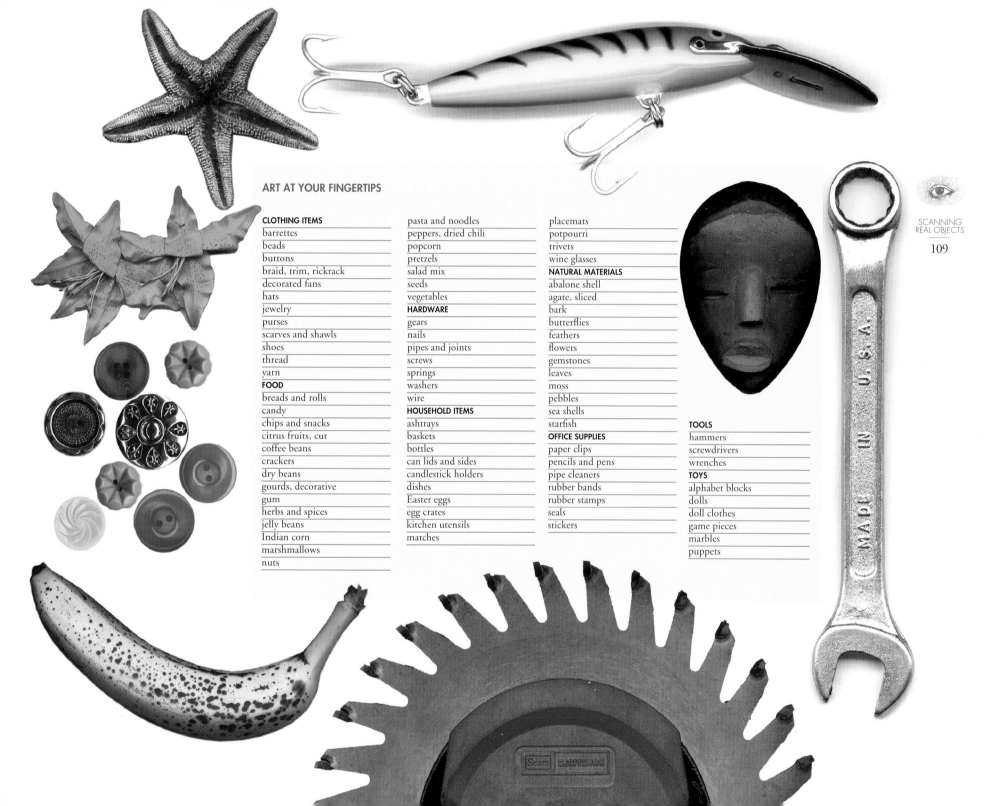

ART AT YOUR FINGERTIPS

CLOTHING ITEMS
barrettes
beads
buttons
braid, trim, rickrack
decorated fans
hats
jewelry
purses
scarves and shawls
shoes
thread
yarn

FOOD
breads and rolls
candy
chips and snacks
citrus fruits, cut
coffee beans
crackers
dry beans
gourds, decorative
gum
herbs and spices
jelly beans
Indian corn
marshmallows
nuts

pasta and noodles
peppers, dried chili
popcorn
pretzels
salad mix
seeds
vegetables

HARDWARE
gears
nails
pipes and joints
screws
springs
washers
wire

HOUSEHOLD ITEMS
ashtrays
baskets
bottles
can lids and sides
candlestick holders
dishes
Easter eggs
egg crates
kitchen utensils
matches

placemats
potpourri
trivets
wine glasses

NATURAL MATERIALS
abalone shell
agate, sliced
bark
butterflies
feathers
flowers
gemstones
leaves
moss
pebbles
sea shells
starfish

OFFICE SUPPLIES
paper clips
pencils and pens
pipe cleaners
rubber bands
rubber stamps
seals
stickers

TOOLS
hammers
screwdrivers
wrenches

TOYS
alphabet blocks
dolls
doll clothes
game pieces
marbles
puppets

SILHOUETTING SCANNED OBJECTS

The uneven backgrounds and multicolored shadows of scanned objects are often messy and need to be cleaned up. Here are two methods, both done in Photoshop, for removing unwanted backgrounds.

Arrange objects alone or in groups on your scanner and scan at your desired resolution.

METHOD 1

1 Select object with lasso tool
In Photoshop, use the lasso tool to select the object, either by drawing around its edges using the mouse or a stylus, or by holding down the Option key and clicking from point to point. When the object selection is finished, choose Inverse from the Select menu to select the background.

2 Delete background
If the silhouette will be used alone, you might want to apply a feather to the selection to create a softened edge. Then use the Delete key with the background color set to white to delete the background.

If you are planning to add a drop shadow however (see next page), the feathered edge will create a lightened halo around the object when it is placed over the darker shadow. To avoid this, do not apply a feather to the selection and make sure the selection is as close as possible to the edges of the object.

METHOD 2

1 Select background with magic wand
If your object has a very wiggly outline, the magic wand may be an easier and faster selection tool than the lasso. Click on an area of the background with the wand and then adjust the tonal tolerance of the tool or Shift-click to add new colors, until you've picked up as much of the background as possible without spreading the selection into the object itself. We used a tolerance of 40 to select almost all of the gray background and then applied a feather to the selection to smooth rough edges.

2 Delete background
We deleted the background selection and found that the remaining areas of gray created a pleasing shadow. This could be used as-is, or we could have continued to use the wand to select areas of background until it was all deleted.

EXPORTING WITH A CLIPPING PATH

The white background around a silhouetted object will appear opaque when the image is saved as a TIFF and imported into a page layout program. To import only the object, export it from Photoshop with a *clipping path* as follows:
1 Select the object using Method 1, Method 2 or some other means.
2 Choose Make Path from the Paths palette to convert the selection to a PostScript path.
3 Choose Save Path to save the path with a name (Path 1).
4 Choose Clipping Path to save Path 1 as a clipping path.
5 Save the document as an EPS.

Go Big!
Take advantage of the magnifica-
tion capabilities of your scanner
to make small objects really large.
This pen was scanned at 250 dpi
at 200%. Remember that all the
imperfections in an object (for
example, the tooth-marks on this
pen) will be magnified too.

CREATING THOSE NICE SOFT DROP SHADOWS

1 Select object
Click on the
background with
the magic wand,
then choose In-
verse from the
Select menu to
select the object.

2 Save selection
Save the selection
into two channels
(#4 and #5) using
the Save Selection
command.

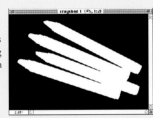

3 Offset and blur
With channel #4
active use the
Offset filter to
offset the image
down and right.
Apply the Gauss-
ian blur filter.

**4 Load original
selection**
Load the selec-
tion in channel
#5 into channel
#4.

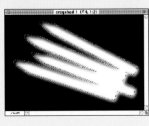

5 Fill with black
Set the back-
ground color to
black and press
the Delete key
to fill the selec-
tion with black.

**6 Load shadow
selection**
Make the RGB
image active and
load the selection
in channel #4.

7 Fill with black
Set the back-
ground color to
black and press
the Delete key to
fill the shadow
selection with
black. Because
the edges of the
selection are
blurred, the edges
of the drop
shadow are
blurred and soft.

CreativityMart

**SALE OF BASIC
SUPPLIES ENDS
SATURDAY!**

Once you've created a clean silhouette of a
scanned object and added a soft drop shadow, it's
ready to be imported into a page layout program,
where it will pop off the page with the imme-
diacy of a good photograph.

Object into Graphic

CREATING A FISH PRINT

This splendid striped bass was bagged on a hazardous expedition to the seafood counter of the local market. The tendency of one-pass scanners to generate spurious colors in metallic objects works to advantage here, producing an intriguing iridescent effect. A white cloth was draped over the fish before scanning with the lid removed. We silhouetted the scanned bass and added a soft drop shadow using the techniques explained in detail on pages 110–111.

A

B

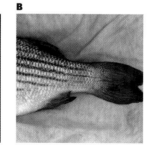

C

Three-dimensional objects that are too large to fit on the scanner can, like flat artwork, be scanned in parts and reassembled in an image-processing program. The canvas size of the file containing the body was increased in width (**A**) to accommodate the tail (**B**). Using a feathered selection the tail was grafted onto the body (**C**) (For more tips on handling large originals see page 14.)

POLARIZING COLORS

Objects that have a metallic sheen, including scaly fish, often pick up multicolored image artifacts when scanned (see "Tips for Scanning Objects" on page 108). To get a realistic color image of a metallic or very shiny object, it's best to photograph it and then scan the photo. But for a special effect, try exaggerating artifacts in an object scan by maximizing the contrast and adjusting the brightness in each of the primary color channels. The two main color systems, RGB and CMYK, give different results, so experiment to find the effect that works best with your image.

Fish processing
A second scan was made with the lid open and the fish draped with black cloth. This image was saved in CMYK mode. Then the black channel was set to 100% contrast and –50% brightness (**A**). Taking the process a step further, we applied the same contrast and brightness changes to the other three channels (**B**).

A

B

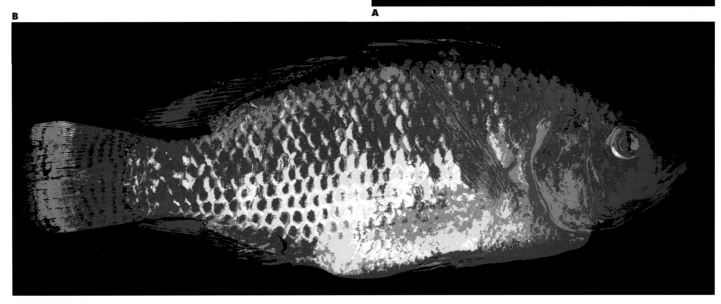

Changing colors
Inverting some of the channels produced colorful variations.

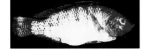

POSTERIZING AN OBJECT SCAN

An image processor can posterize automatically by reducing the tonal values of an image. This works quickly and well in grayscale, but in color the results are often harsh (see "Posterization" on page 88). Another way to produce a good color posterization is to sample one dominant color from the image for each step, edit the density of each step in a separate channel and assign specific colors to each channel.

Posterizing a leaf
A color scan of a leaf (**A**), was sampled in three places to form a separate palette file (**B**). After three new channels were established, the image was copied into each one, inverted and set to 100% contrast. Then brightness was adjusted: −50% brightness (**C**), 0% brightness (**D**) and +85% brightness (**E**). In the color image the selection made from each channel was filled with a color from the palette, building color in layers.

VOLUME 3 NO. 1
WINTER 1993

CALIFORNIA ENERGY
EXTENSION SERVICE

AN ENERGY
MANAGEMENT ACTION
PROGRAM OF THE
OFFICE OF GOVERNOR
PETE WILSON

California Indian *Energy News*

Tribal Perspective

Wood: A Natural Fuel for Resighini Rancheria

"Everyone likes the new heating system; all the rooms are much warmer than they were before the heating retrofit."
—*Donald McCovey*
Tribal Chairman

Resighini Rancheria, 228 acres situated on the lower Klamath River is home to 53 people of Yurok ancestry. Alder trees stretch upward to hills blanketed in conifers. Wood is a natural fuel for Resighini residents. A recent retrofit of the tribal building's heating system enables the Tribe to burn their wood more efficiently.

Heating system retrofit
The previous heating system for the tribal building was a diesel-fired hydronic system. The system was used until the diesel fuel tank ran dry and then, between fuelings, an old Franklin wood stove heated the building. Baseboard heaters distributed heat to the

offices and the main meeting room. The old wood stove needed constant feeding of wood fuel and at best only heated the space in close proximity to the stove. The efficiency of the old stove was estimated at 40 percent or less. The smoke and particulate matter that leaked into the room and the amount exhausted outside were excessive when compared to today's more efficient wood stoves. After receiving a request from the Tribe, CEES asked contractor Sequoia Technical Services (STS) to design an upgraded heating system. STS recommended that the Tribe disconnect the diesel heater and purchase a

continued on page 10

A NEWSLETTER COVER FROM A SCAN OF A LOG

The flat end of a large log perched on the scanner produced a passable grayscale scan, but posterizing brought out the detail of the wood grain and gave a graphic texture to a cover design. The automatic posterizing menu was set to three levels.

It takes a steady hand

The best results from scanning can be obtained by holding the hand an inch above the glass. Contacting the glass produces a squashed appearance. Jagged edges that result from minute motions of the hand can be edited out in silhouetting.

A PORTRAIT FROM SCANNED BODY PARTS

Few scanner owners we know have been able to resist the temptation to scan some part of their anatomy, but we suspect that few have come up with results that have worked their way onto the printed page. It's worth persevering and experimenting, however, because the images can be interesting, and the model fees are really affordable.

Scanning faces

Flatbed scanners are optimized for two-dimensional originals and were not designed to scan large, irregularly shaped objects, such as heads. Because the scanner's depth of focus is limited, profiles work better then full face (unless you have an unusually small or flat nose). Distortion often occurs (**A**) but can easily be corrected by stretching the image. If color balance is a problem, scan in grayscale (**B**). Silhouettes (**C**) can also be effective. The hand in this line scan was placed in contact with the glass and covered with a black cloth. (**D**) The grayscale profile and the hand silhouette was combined, converted to a dithered mezzotint image, and montaged with other dithered elements for a poster design (**E**).

A

B

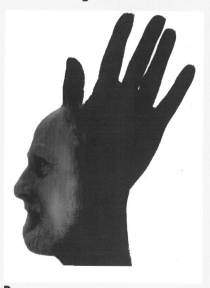

C

D

E

A MONTAGE OF OBJECT SCANS

It's not difficult to find and put together objects that will make up an interesting montage. Unlike still-life photography, in which each object is in realistic proportion with other objects, a montage need not limited by the relative scale of the objects, providing more options in composition.

We used a thematic and allegorical approach with the montage on this page; objects are not literal but represent other things. Another approach might be completely abstract, in which shape, color and composition predominate, and the meaning of the objects is irrelevant.

A

B

Silhouetting
Each element of the montage is a subassembly. Murky backgrounds were deleted by feathered selection (**A**) or by adjusting the tonal levels (**B**).

Computers
in an age
of innocence

A study of the impact of new technology

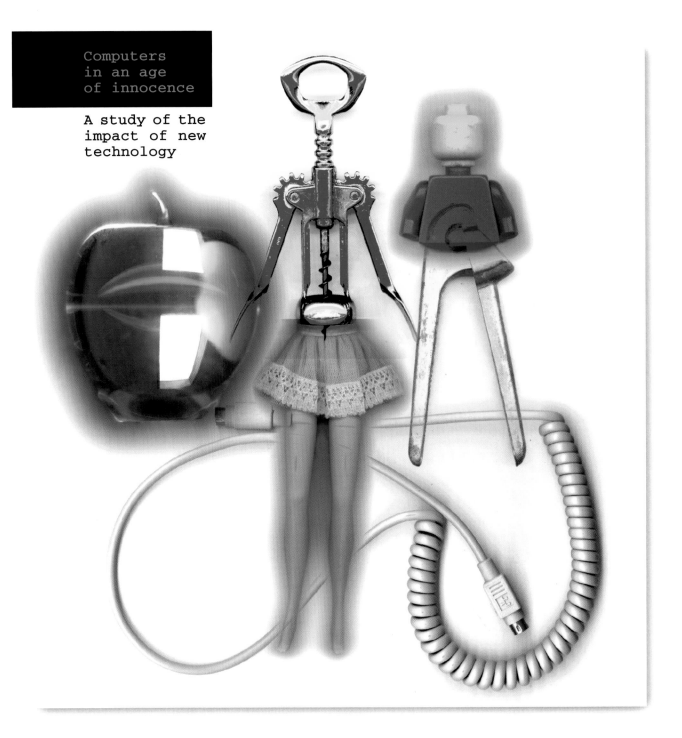

12 | Creating 3D Illustrations

Extruding Scans in 3D

FROM 2D TO 3D

Almost any line art image can be imported and converted into a three-dimensional shape in a desktop 3D program. Most programs accept bitmap scans, but some may require the image to be autotraced and turned into a PostScript graphic first.

When a shape is extruded, its two-dimensional outlines are extended in depth, rather like a loaf of bread arising out of a repetition of a single slice. An extruded shape looks no different from its flat progenitor unless viewed at an angle, but you may not like the distortion the perspective brings. One solution is to bevel the edge of the object while it is extruded. The beveled edge catches the light and brings an illusion of depth without distorting the object. If your program gives you control over the amount of bevel, be subtle!

Beveled edges
A crisply stylized hawk (**A**) scanned from *Visual Elements 1* (Rockport Publishers, 1989) was first inverted (**B**), then outlined with a heavy stroke (**C**). Once imported into a 3D program it could be extruded and beveled. Beveling gives edge detail and increases the sense of depth. The finished rendering (**D**) has a brushed chrome finish applied and an extra light source to bring out the sheen of the metal surface. The purplish highlights are the result of a "chrome" environment map.

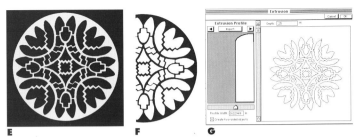

E   F   G

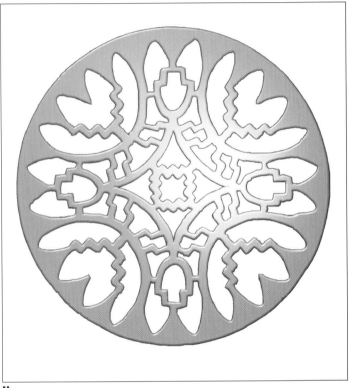

H

A

B

C

D

I

Extruding a Decorative Shape
We started with an ornamental design from East Anglia, A.D. 625, scanned from *Early Medieval Designs from Britain* by Eva Wilson (Dover, 1983) (**E**). It was inverted and imported into a 3D program (**F**). The imported image was transformed into an object that could be extruded (expanded in depth) and given a beveled edge (**G**). A texture of gold sheen applied to the surface gave a realistic appearance to the final rendering (**H**). We incorporated the same shape into a 3D model of a woman's head as an earring (**I**). Note the shadow of the earring cast on the shoulder of the jacket.

A

A CITYSCAPE FROM EXTRUDED CLIP ART

In a curious realm that's somewhere between realism and a cartoon, the 3D program's camera roams a rather smoggy and noticeably underpopulated metropolis. To make it work, we took some 1930s chart icons (**A**) from the *Handbook of Pictorial Symbols*, edited by Rudolf Modley (Dover, 1976), and extruded them upon a simple grid of city blocks (**B**). The buildings were moved into position and a camera was placed in the scene (**C**). After some test shots with the camera, we decided to add more city on either side (**D**), easily done by grouping the model and replicating it twice. The finished rendering (**E**) required ray-tracing, which gives accurate shadows and reflections, and a fog filter to lend an eerie presence.

E

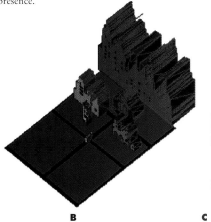

B

C

D

Scans as Texture Maps

IMAGE MAPPING: AN OVERVIEW

In 3D programs any surface or shape can be assigned attributes that affect how it will appear when rendered. In the texture palette of many 3D programs there are channels that can "map" scanned images to modify the surface characteristics of the object in various ways.

An object may be a solid color or, by placing an image of a texture into a color map channel—for instance, a scan of stone—you can cover the object's surface with a repetition of that image. Other kinds of maps can be applied via different channels including bump, glow, reflectance and transparency. A reflectance map, for example, governs the amount of glossiness over specific areas of an object.

Texture map
A scan of a photo of exposed aggregate concrete was mapped directly onto the surface of the monolith. This is the simplest form of texture mapping.

A TRANSPARENCY MAP OF SATURN'S RINGS

One could draw each of the myriad rings as a separate object and then assign them all different colors, but a better way is to use a texture map to project an image of the rings onto a single object.

In Strata StudioPro we put the color image in both the Specular Color and Diffuse Color channels. Specular and Diffuse channels contain the negative grayscale version of the image. Transparency contains the grayscale version of the image. Reflectivity and Glow were left empty.

Bump mapping
A bump map distorts the surface to which it is applied by raising the light areas and lowering the dark areas. So, a scanned photograph of actual ripples makes an ideal bump map for simulating the surface of water. If the surface of the water texture is reflective and the lighting and camera angles are considered carefully, the effect of a shimmering pool is dramatic.

Using scanned textures in 3D models
This image of a monument after a rain shower combines texture, bump and reflectance mapping. The foliage on the trees was made by applying a texture map to a semitransparent surface. This is not quite the same as a true transparency map. (See the sidebar on the rings of Saturn on this page).

Reflectance mapping
To the texture map of concrete (above) we added this pattern —a detail from the sponge texture shown on page 60—as a reflectance map. The texture was given the overall attribute of mirrorlike reflectance, but the reflectance map allows only those areas that are white to be reflective and suppresses reflectiveness in the black areas. Hence the puddles.

View through a virtual camera
In this "wireframe" screen representation in StudioPro (**A**) an invisible light source is shown as a giant light bulb. The faceted surface of the lime will be smoothed out in the final rendering.

A

ADDING REALISM TO A MODEL

By applying scanned textures from the real world to the surfaces of a model in a 3D rendering program, you can achieve a dreamlike hyper-realism similar to that found in the paintings of Magritte. Although the shapes within the model may be simple, they come to life when "mapped" with realistic textures. Color texture images, like wood or rock, are called *texture maps*; grayscale images that add depth detail to a surface, such us ripples in water, are known as *bump maps*. Texture and bump maps are combined with specifications for color, glossiness and opacity to produce surface textures that are startlingly realistic. We used these techniques to place an oversize lime inside a wallpapered room.

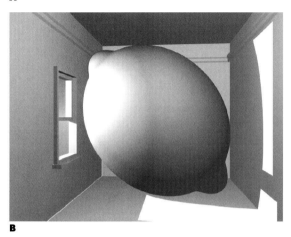

B

Test rendering without any textures applied
Lighting is the key to successful 3D images and should be tested before time-consuming textured renderings are attempted. in this model, an unseen light illuminates the interior but casts no shadow; another light, hidden from view outside the window, casts dramatic shadows on the opposite wall (**B**).

A texture map from a found object
We took a slice of lime skin, about 3 inches square, flattened it out and placed it on a flatbed scanner (**C**). Then in Photoshop we used the Offset filter to offset the image down and to the right and chose Wrap Around as the method for filling the Undefined Areas so that the parts of the image that were pushed off the bottom and right edges were wrapped around to the upper and left edges of the image area (**D**). Using the rubber stamp tool, we disguised the splices to form a seamless tile (**E**). To add further realism we also made a bump map to mimic the pitted surface of the fruit (**F**). This was done by finding the CMYK channel that had the most texture detail (in this case, yellow), saving it to a grayscale file and inverting the image.

The scale and orientation of the texture as it is applied to the object can be altered. Numerous test renderings were necessary to find the correct settings (**G**).

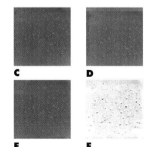

C　　**D**

E　　**F**

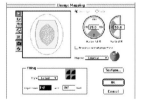

G

Making the floor and walls
To create floor boards, a photograph of decking (see page 70) was applied to the floor surface as a texture map and given a coat of varnish by altering the color and adding reflectivity (**H**). A wallpaper sample from an old catalog was mapped onto the walls. The image was cropped as shown (**I**) to make the pattern repeat.

H

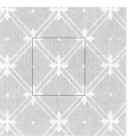

I

A MOCK-UP OF A PACKAGE DESIGN

Most 3D rendering programs produce bit-mapped images, but object-oriented 3D programs, such as Adobe Dimensions, produce renderings in PostScript format which can be output at any size without loss of resolution. This approach is ideal for handling projects such as package designs that involve graphic elements whose integrity must be preserved through any amount of scaling or distortion of perspective.

The starting point for our box design was a grayscale scan of desert mountains (**A**). With a halftone filter applied (**B**), the image was broken into horizontal lines (see page 98), which were then autotraced in a PostScript program and combined with typography (**C**).

A

Next we drew a simple box in Dimensions and imported the artwork into the Artwork Mapping window (**D**). The shaded portions show the parts not visible from the selected viewpoint. After checking a keyline preview (**E**), the box color was specified in the Surface Properties window (**F**). We adjusted the lighting to control the relative tone of the three visible sides (**G**).

B

Women's Size 12 **All-Terrain Boots**

C

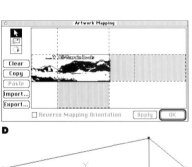

D

E

F

G

13 | Multimedia Projects

Designing for the Screen

THE SCREEN VERSUS THE PAGE

Most of the art we've included in this book is intended for print media. But as electronic media become more widespread, illustrators and designers find themselves being asked to design everything from simple on-screen slide shows to Web page graphics, to complex interactive CD-ROM "books," or even 30-frame-per-second, full-screen animation. Scans, used with attention to fundamental design principles, can be very helpful in designing for the screen.

On-screen presentations benefit greatly from the traditional skills and concerns that designers have always brought to graphic design; for example, balanced composition, readability and coherence. Sadly, we've noticed that some interface designs are so densely filled with spinning 3D spheres, marbled backgrounds, and extruded metallic surfaces that they look like science fiction spaceship controls created by interior decorators gone wild. Just because a special effect *can* be done on the computer doesn't mean it *should* be done.

Yes, there are fundamental differences between the page and the screen, but we tend to approach on-screen design with the same esthetic and as much restraint as we do the printed page. In other words, we appreciate white space, an absence of clutter and elegant simplicity. These values are especially important for on-screen graphics because the screen simply cannot support as much "information density" as the printed page can. For one thing, the simple presence of a screen seems to make the adult viewer expect to be passively entertained, rather than actively engaged in a learning process. In addition, viewing type and small images on an electronic monitor is harder on the eyes and brain because of the flicker of the screen and also because the light emitted is greater than that reflected by a piece of paper. These factors are especially important when it comes to type.

MAKING TYPE READABLE

We have found that it's just not comfortable to read a lot of type on-screen, primarily because it appears at about 1/35th of the resolution of a typical printed page (72 ppi on-screen as opposed to 2400 dpi for high-quality printed type). To adapt, on-screen type must be set larger and with more leading and should have plenty of letterspace. It's also important to experiment with different typefaces to find the ones that are most readable on-screen. Choose type without a lot of contrast between thick and thin strokes and don't use script fonts, ornamental italic, or all caps. We've found that simple sans serif faces (Futura, Helvetica, Franklin Gothic and so on) work best for blocks of smaller type, whereas serif faces can be used as heads at larger sizes. Also, be sure to place type over a simple, contrasting background.

PRODUCING A TACTILE LOOK

Another difference between the printed page and the screen is that screens are usually interactive; that is, they contain navigational buttons and other elements that must be "touched" with the mouse cursor so that the viewer can move from one screen to another within the presentation. Screen designers often make these buttons look more "pushable" by using 3D effects or by stylizing the "hot" areas of the screen in some way that sets them off. The goal is to make the screen self-explanatory and intuitive by giving users simple or conventional visual cues that guide the eye and hand to the right areas without too much thought or searching.

CREATING A SENSE OF DEPTH

Another fascinating difference between the screen and the page is that even though both are flat surfaces, the screen seems to contain more depth. We have found that it's very rewarding to create on-screen images that contain layers of elements and to provide soft drop shadows behind them, so that elements appear to float over each other in a three-dimensional space. Adding depth helps compensate for the lower resolution of the screen by providing a richness of representation similar to the pseudorealism of a *trompe-l'oeil* painting. (For techniques for producing drop shadows see page 111.)

RESOLUTION, SCREEN SIZE AND COLOR DEPTH

Determining the correct resolution for multimedia is easy: 72 ppi is the standard for on-screen displays, whether for CD-ROM presentations, Web pages or output for video. And no matter what the size of the viewing monitor, the proportions of a multimedia image are the same: 640 pixels wide by 480 pixels high for CD-ROM, with the same ratio used when images are enlarged on a video monitor.

RECOMMENDED FORMATS FOR MULTIMEDIA IMAGES
(MACINTOSH PLATFORM)

MEDIUM	SOURCE FILE	COLORS	WIDTH	HEIGHT
Slide show	PICT	16 million	640 pixels	480 pixels
CD-ROM	PICT	256	640 pixels	480 pixels
Photo CD (Kodak)	PICT	256	640 pixels	480 pixels
Web	GIF	256	475 pixels	300 pixels
Video clip (Premiere)	QuickTime	256	320 pixels	240 pixels
Animation (Macromedia Director)	PICT	256	512 pixels	314 pixels
Animation	QuickTime	16 million	640 pixels	480 pixels

Recommended screen format for multimedia

We created this template in a PostScript program and use it as a universal grid for multimedia projects. All dimensions are given in inches. The inner gray area is the "safe" area, where all images will be seen by all monitors regardless of age. The outer yellow rectangle corresponds in inches to the normal 640 × 480 pixels at 72 ppi. The outermost rectangle is for NTSC video applications where possible monitor misalignment must be accommodated.

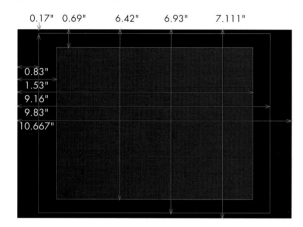

When creating graphics for multimedia, it's best to view your image at actual size and periodically switch from 24-bit (millions of colors) to 8-bit (256 colors) color mode, since your graphic should look good even on a low-end monitor. Save your files in both 24-bit and 8-bit versions, since your final conversion to 256 colors will be irreversible. Note that areas of solid color, such as panels behind lettering, should be filled with colors chosen from the System palette, since other hues may develop unwanted texture patterns when dithered.

Some multimedia projects may require a special palette that has been designed to optimize video, to accommodate users with less than 256 color systems, or to work with the Windows palette, which is somewhat different than the Macintosh System palette. This requirement will not affect the initial scanning, which should always be done at 16 million colors, but it's a good idea to see how the images will look when indexed in a limited palette, in order to make adjustments to the color if necessary. If your image is to appear in video, use 24-bit color. Note that some colors, especially yellows, oranges and reds, look garish on the video screen and may need to be edited.

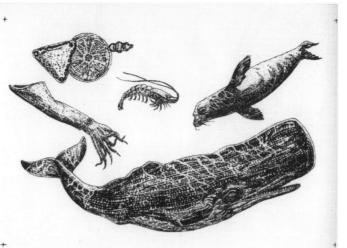

Using scanned images in CD-ROM

These screens are part of a prototype for an educational CD-ROM on ecology. The line art for the marine animals was created by Karl Nicholasen in Colorado and faxed to California to meet a tight deadline. We scanned the faxed copies and were able to successfully incorporate them into the images because of the low screen resolution of 72 ppi.

To assist the student users of the CD-ROM, the interactional buttons were placed on the "top" or most forward layer of the dimensional hierarchy, while the images to be viewed are in the background layers. Most of the graphic content of the screens was derived from scanned material.

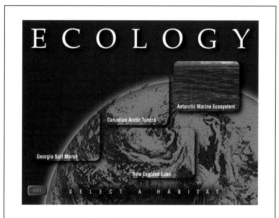

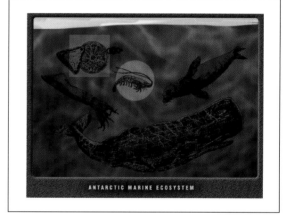

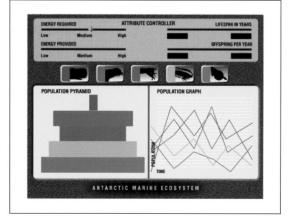

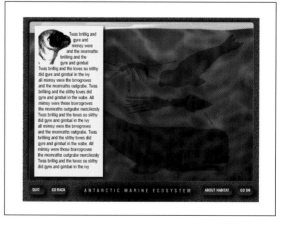

Designing for Multimedia

CREATING SCREENS FOR CD-ROM

The great thing about designing for CD-ROM is that what you see is indeed what you get. It's much easier to predict how your finished image will look on another user's screen, even given the variance between monitors, than to try to anticipate whether your design will translate well when printed with ink on paper.

At this point, the technology supporting graphics for CD-ROM presentations is more advanced than the technology for the Internet (see page 128) so it's possible, for example, to include many graphics without being concerned about downloading time for the user. Also, the type specifications and design formatting you create on your CD-ROM page will be preserved in the finished PICT images and will not be affected by the user's viewing system, which is not the case for Internet graphics. CD-ROM presentations can also include sound and partial-screen animation or video clips.

WORKING IN LAYERS

Because multimedia projects are often group efforts, created by many designers, artists, writers and editors in different locations, the elements of a particular presentation can (and probably will) change while the work is in progress. We have found that the Layers function in Photoshop is ideal for organizing the scans and other elements that make up multimedia screens. It makes it possible to assign elements of different types—buttons, button titles, headlines, title bars, background graphics, foreground graphics, blocks of type and so on—to different layers, so that late-breaking changes from editors or art directors can easily be made to one or two layers without affecting the graphic content of the other layers. When final approval has been given, a copy of the layered Photoshop document can be collapsed and saved as a conventional PICT.

BUILDING INTERACTIVE SCREENS IN LAYERS

Creating interactive screens in layers in Photoshop makes it possible to edit single elements without affecting the rest of the illustration. To start a prototype screen for *Peter Norton's PC Guru* (Verbum, 1996), a scan of a spiral notebook was placed in the background layer. A scanned silhouette with drop shadow and color bars were placed in layers above (**A**). Next, objects relating to the topic were scanned and placed in separate layers above the background (**B**). Label buttons for the items were positioned (**C**), a title and running head were created (**D**), and a navigational ball and buttons were added to complete the screen (**E**). The Photoshop layers palette shows how the layers were named and positioned in the hierarchy (**F**).

CREATING AN UNDERLYING GRID

Designers are often called upon not only to create pleasing graphics for a CD-ROM presentation but to help plan the flow of the presentation and the elements that make it work. We have found that it's helpful to create a prototypical screen in a PostScript drawing program—a simple lines-and-boxes layout that includes the screen borders, background panels, buttons, headlines and so on—and import this into Photoshop as a background layer to help with aligning elements so that all screens for a project follow the same format.

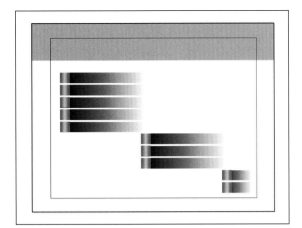

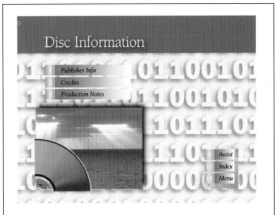

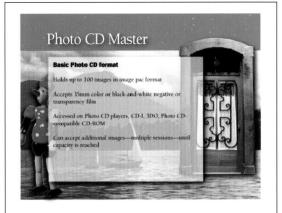

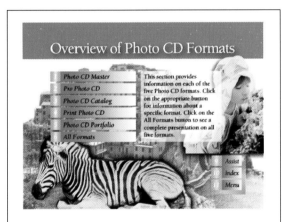

Screens for Photo CD Portfolio

These screens, created for a Photo CD Portfolio presentation, share a common architecture of button location, banner treatment, color palette and typography, as indicated by the schematic (above left), which was created in Illustrator and opened in Photoshop to serve as a template in a background layer. The visual elements of the screens derive primarily from scanned material, including original photographs, marbled paper and stock photo elements provided on Kodak Photo CD.

To make type elements more readable, while still preserving the background graphics, translucent panels were floated over the images to provide a lighter-colored panel for type.

This informative CD about the different types of Kodak Photo CD is available with *The Official Photo CD Handbook*, by Michael Gosney (Peachpit Press, 1995) and includes original music composed and performed by Janet Ashford.

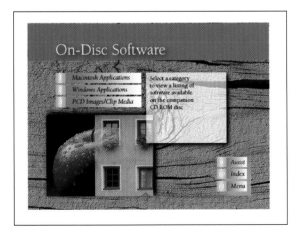

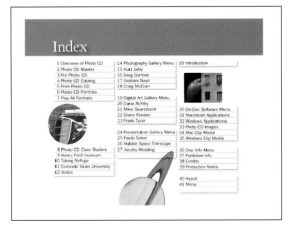

CREATING GRAPHICS
FOR THE WORLD WIDE WEB

The World Wide Web is an area of the Internet in which "home pages" posted by various companies, institutions, and individuals can be viewed and linked with each other. Web pages at their simplest consist of windows containing text, but increasingly include graphics and sound files designed to make presentations more appealing and accessible. Web pages are typically defined in a computer language called *HTML* (*hyper text markup language*). The graphics and text of Web pages are actually transmitted via programming code that defines positioning of elements as well as content, and that enables links, so that clicking on a highlighted area on a Web page will move the user to another domain in the same home page, or to another home page somewhere else on the Web.

Designing pages for the Web presents many challenges. When using HTML, the designer can determine only some of the elements of the page (the imbedded graphic images, for example). Other elements, particularly the typeface, color and size used for the running text and the color of the background, are determined by the Web browser and by specifications defined by the system displaying the page. In addition, the width of text columns can be enlarged at will by the viewer by resizing the window. Since text blocks rewrap to fill the new width, this can lead to the appearance of wide, unreadable lines of type.

Layout and typography specifications can be preserved by saving a web design through Adobe Acrobat as a *PDF* (*PORTABLE document format*) file. Then the design will appear on the Web looking exactly the same as it did on the designer's screen. For more information on PDF contact Adobe Systems (see Resources on page 133).

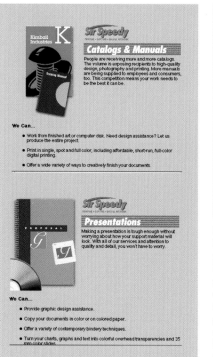

Matching the background
Most web browsers define a 20 percent black screen as the background for Web pages. If you want your irregularly shaped graphics to look like they're floating over this background (rather than appear in a white rectangular bounding box), be sure to fill the background of your graphics with a 20 percent black fill.

Starting with a full-page graphic
A typical home page for the World Wide Web opens with a full-screen graphic that makes a good visual first impression, identifies the owner of the site and serves as a table of contents. For the Sir Speedy Web page, John Odam created a graphic that contains title buttons for the different services offered. Clicking on a button links the user to a *domain screen* describing that service.

Designing domain screens
We used a page layout program to work out the positioning of the elements on all the domain pages. The HTML programmers used the layout files as a guide to formatting the text.

THE DIGITAL WELCOME MAT

Because it often takes a long time for online images to form on the screen, full-screen graphics are usually used only as the opening page or *home page* of a Web site, providing a master image with an array of contents buttons. The visual impact of a home page should be worth waiting for, as shown by the examples on this page.

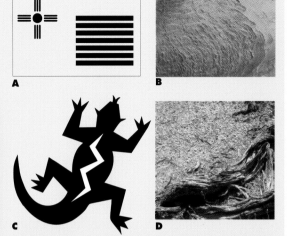
A

B

C   **D**

Project for an online magazine
These designs arise from the default gray background of the browser and achieve a sense of depth with shadows. By convention, a blue box indicates a button.

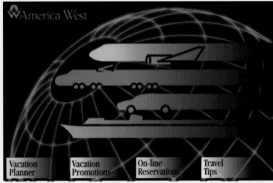

Online travel
This home page for an airline uses only visual icons as buttons to take you to four domains: flights, train services, car rental and cruises. The fade-out stripes suggest motion.

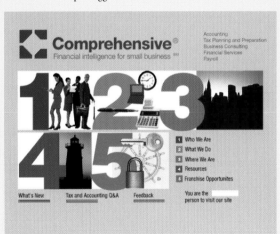

Warm numbers
Large numbers in welcoming colors are the main emphasis of this home page for an accounting and financial services company.

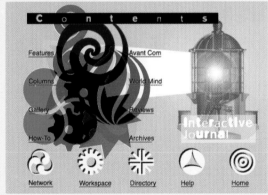
E

Global village government
Politicians have been quick to realize the public relations potential of the Web. It offers them an inexpensive way to get their message across and get feedback from their constituents. The basis of this design is a simple symbol and set of bars (**A**) drawn in a PostScript program and embossed into a texture scan of sandstone (**B**). For a Southwestern touch, our lizard (**C**), adapted from an icon font (*DF Naturals* from Fontex), was filled with sand texture (**D**), and placed with a soft shadow onto the sandstone. We added type, ensuring that the background was always dark enough for the white lettering to show clearly (**E**).

A

B

C

D

E

CREATING GRAPHICS
FOR ANIMATION

When desktop scanners were first introduced, their application for illustration, photo-editing and design was immediately apparent. But as scanners have evolved along with the development of desktop 3D and animation programs, scanning has become a source of cell graphics for animations that are either viewed on-screen or output to video.

The key to successful animation, especially for beginners, is to keep things simple. In fact, your first animation project should probably be a "just for fun" experiment like many of those shown on these pages.

SIZE AND FILE CONSTRAINTS

All images used in animation programs must be in PICT format, rather than TIFF. The size and color depth of graphics used for animations should be adapted to the end purpose. In general, if your animation will be viewed on the computer screen only, create your graphics at 72 ppi (screen resolution) at actual viewing size in 8-bit color. For animations that will be output to video, the screen size is limited to 640 × 480 pixels at 72 ppi, the same format as for CD-ROM screens, but color depth can be increased to 24-bit.

A minimalist animation
Animating an entire illustration is time-consuming. You can save time and get almost the same effect by having only one element of an image move. This man was created by drawing with a fine-tip marker on bond paper, at about postage stamp size. Scanning at 400 percent enlarged the drawing and exaggerated the pen markings to create a warm feeling. Three versions of the mouth were created—closed (**A**), open (**B**) and vowel position (**C**)—and were sequenced in Macromedia Director, along with some type (**D**, **E**) to match the audio voice-over, "You really *like* this design and you *will* approve it."

A

B

C

Study for an animated logo
To create a moving logo for a fictitious travel company, John Odam started with scans of a 20-frame photo sequence (see sidebar). Each black-and-white photo was scanned separately and placed into the "cast" window (**A**) of a Macromedia Director document. He added color in Director so that the elephant changes hue in a psychedelic way as it moves. A copy of the same sequence was placed in another channel of the animation hierarchy and offset by a few frames, so that there appear to be two elephants walking (**B**, **C**), as the title flashes on and off.

MUYBRIDGE'S IMAGES:
A SCANNING SOURCE FOR ANIMATION

Turn-of-the-century photographer Eadweard Muybridge was fascinated with locomotion and made hundreds of sequential photographs of people and animals walking and running. The photos, reprinted by Dover, can be scanned and used as the basis of charming animation sequences as well as still art.

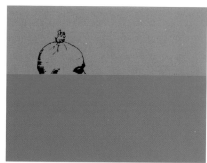

Doing it the old way: one drawing per frame
In the early days of film animation, artists created a separate drawing for each frame of the film, generating thousands of drawings to produce a few minutes of motion. We wouldn't attempt this on the desktop, but just for fun, we scanned a series of very small line drawings, which had originally been made by artist Karl Nicholasen as a flip-book animation for the corners of the pages of a psychology textbook. Each small drawing was scanned as line art at 72 ppi at 400 percent and placed over a colored background layer in Macromedia Director. The horizontal line formed by the boundary of the blue and green areas in the background served as a reference for registering the line art.

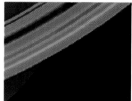

Generating frames with a 3D program
In addition to generating three-dimensional objects, 3D programs also include cameras that can be moved through a 3D model to take a sequence of "pictures" that simulate a fly-through animation. The technique differs from a simple "zoom" or enlargement of an object because as the camera moves it captures the model from a slightly different perspective each time, adding an uncanny sense of depth and motion. The camera can be made to fly through the model (go toward the object, as shown above) or pull back (as shown at right).

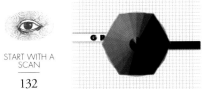

A moving montage in layers

This storyboard outlines a 60-second animation loop created to run in kiosks for a chain of instant-print franchises. The animation describes the six services available at each store: design, printing, copying, binding, digital services and mailing, each of which is featured for 10 seconds. The frames combine scans of line art and real objects with 3D elements. The 640 × 480-pixel screens were assembled in Macromedia Director and transferred to video tape.

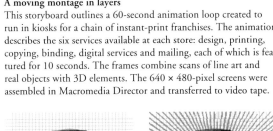

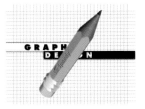

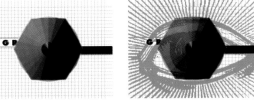

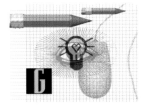

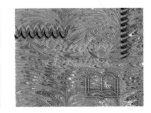

Resources

PRINTED CLIP ART

Art Direction Book Company
10 East 39th Street
New York, NY 10016
212/889-6500

Crown Publishers
225 Park Avenue South
New York 10003

Dover Publications
31 East 2nd Street
Mineola, NY 11501

Fontex
Letraset USA
40 Eisenhower Drive
Paramus, NJ 07653

Rockport Publishers
5 Smith Street
Rockport, MA 01966
508/546-9590

Stemmer House Publishers, Inc.
2627 Caves Road
Owings Mills, MD 21117-2998

CD-ROM IMAGES

Artbeats Software, Inc.
P.O. Box 709
Myrtle Creek, OR 97457
800/444-9392, 503/863-4429
503/863-4547 fax

Classic PIO Partners
87 East Green Street, Suite 309
Pasadena, CA 91105
800/370-2746, 818/564-8106

CMCD c/o PhotoDisc, Inc.
2013 Fourth Avenue, Suite 402
Seattle, WA 98121
800/664-2623, 206/441-9355
206/441-9379 fax

Digital Media Corp.
28362 Via Dandina
Laguna Niguel, CA 92656
800/786-2512, 714/362-5103
714/642-2426 fax

Fractal Design Corp.
P.O. Box 2380 Aptos, CA 95001
408/688-5300
Maker of: Painter

Gazelle Technologies, Inc.
7434 Trade Street
San Diego, CA 92121
800/843-9497
619/693-4030
619/536-2345 fax

Image Club Graphics
10545 West Donges Court
Milwaukee, WI 53224-9967
800/661-9410, 403/262-8008
403/261-7013 fax

SOFTWARE

Adobe Systems, Inc.
P.O. Box 7900
Mountain View, CA 94039-7900
800/833-6687
Maker of: Dimensions
Gallery Effects (Vols. 1, 2 and 3)
Illustrator
PageMaker
Photoshop
PhotoStyler
Premiere
Streamline

Altsys Corporation
269 Renner Parkway
Richardson, TX 75080
214/680-2060
214/680-2060 fax
Maker of: Fontographer

Gryphon Software
7220 Trade Street 120
San Diego, CA 92121
619/536-8815
Maker of: Morph

HSC Software
6303 Carpinteria Avenue
Carpinteria, CA 93013
805/566-6200
805/566-6385
Makers of: Kai's Power Tools

Light Source
17 East Sir Francis Drake Blvd.,
Suite 100
Larkspur, CA 94939
415/461-8000
Maker of: Ofoto

Macromedia, Inc.
600 Townsend Street
San Francisco, CA 94103
800/438-5080
Maker of: FreeHand, Director

Quark, Inc.
1800 Grant Street
Denver, CO 80203
800/788-7835
303/343-2086
Maker of: QuarkXPress

Specular International
479 West Street
Amherst, MA 01002
413/253-3100
413/253-0540 fax
Maker of: Collage

Strata Incorporated
2 West St. George Blvd.
Ancestor Square, Suite 2100
St. George, UT 84770
801/628-5218
801/628-9756 fax
Maker of: StudioPro

Treacyfaces, Inc.
P.O. Box 26036
West Haven, CT 06516
800/800-6805, 203/389-7037
203/389-7039 fax
Maker of: Bistro font

Xaos Tools, Inc.
600 Townsend Street, Suite 270E
San Francisco, CA 94103
415/487-7000
Maker of: Paint Alchemy

BOOKS

Legal Guide for the Visual Artist
by Tad Crawford
from Allworth Press
10 East 23rd Street
New York, NY 10010

Real World Scanning and Halftones
by David Blatner and Steve Roth
from Peachpit Press
2414 Sixth Street
Berkeley, CA 94710
800/283-9444

SCANNERS AND OTHER RESOURCES

Eastman Kodak
343 State Street
Rochester, NY 14650-0811
800/242-2424

LaCie
8700 SW Creekside Place
Beaverton OR 97005
503/520-9000

Microtek Lab, Inc.
3715 Doolittle Drive
Redondo Beach, CA 90278
310/297-5000

Index

About the Authors

JANET ASHFORD is a freelance writer and designer and the co-author of four books on computer graphics: *Start with a Scan: A Guide to Transforming Scanned Photos and Objects into High Quality Art* (Peachpit, 1996), *Adobe Illustrator: A Visual Guide for the Mac* (Graphic-Sha/Addison-Wesley, 1995), *Aldus PageMaker: A Visual Guide for the Mac* (Graphic-Sha/Addison-Wesley, 1994), and *The Verbum Book of PostScript Illustration* (M&T Books, 1990).

Over the past eight years Ashford has written regular how-to articles on computer graphics for *MacUser, MacWorld, Step-By-Step Electronic Design, Print, Step-By-Step Graphics, and Dynamic Graphics*. She has created designs for books, newsletters, brochures, and CD packaging, and has produced original illustrations for posters, textbooks, and magazines.

Ashford has also worked as a fine artist for the past thirty years, creating drawings, paintings, posters, and artist books with watercolor, pen-and-ink, oils, acrylics, and silk screen. She is also a musician, and she composed and performed the original music for the interactive Photo CD that accompanies *The Official Photo CD Handbook* (Peachpit Press, 1995). Ashford has played Celtic fiddle with Lime in the Harp and plays early California music with Los Californios.

Before becoming involved with computer graphics, Ashford wrote many books and articles on childbirth and women's, health, including *The Whole Birth Catalog* (Crossing Press, 1983) and *Birth Stories: The Experience Remembered* (Crossing Press. 1984). From 1979 to 1988 she edited and published *Childbirth Alternative Quarterly*, which is archived at the National Library of Congress. Ashford's satirical short story, *Natural Love,* is included in *Cyborg Babies: From Techno-Sex to Techno-Tots* (Routledge, 1998). Her latest childbirth project is a documentary video, *The Timeless Way: A History of Birth from Ancient to Modern Times* (InJoy Videos, 1998). Ashford has a B.A. in psychology from UCLA. She and her youngest daughter live in the town of Mendocino, on the coast in Northern California. Information on all her work, including computer graphics, music, childbirth and psychology, can be seen at www.jashford.com.

JOHN ODAM is an award-winning graphic designer and the principal of John Odam Design Associates. John grew up in England, but left there just as the Beatles were starting to get famous. In art school he mainly learned how to play the guitar and how not to leave his brushes tip-down in the water jar. Unlike most of his cohort, John wasn't yet cynical enough to make a living in advertising, so he gravitated toward publishing.

He was for many years art director of the early desktop-published *Verbum* magazine and now writes regularly for *Step-By-Step Electronic Design* and *Before & After*. He is a co-author of *The Gray Book* (Ventana Press, 1990), a book on techniques for black-and-white computer art. He has contributed many articles and illustrations to magazines and books throughout the world.

John runs a full-service design business producing cover and page designs for college textbooks and trade books, advertising, signage, packaging, catalogs and multimedia. With the advent of online graphics, John's studio is kept busy with home pages for the World Wide Web, as well as CD-ROM screens and animation projects.

John loves to design, and loves unscrewing broken scanners to find out how they work.

Janet and John have worked together for the past ten years, collaborating on various book designs, illustration work and multimedia interface designs.